Climate Change 1992

The Supplementary Report to the IPCC Scientific Assessment

Edited by

J.T.HOUGHTON, B.A.CALLANDER and S.K.VARNEY

Published for the
Intergovernmental Panel on Climate Change

300623487X

Published by the Press Syndicate of the University of Cambridge
The Pitt Building, Trumpington Street, Cambridge CB2 1RP
40 West 20th Street, New York, NY 10011-4211, USA
10 Stamford Road, Oakleigh, Victoria 3166, Australia

First published 1992

Printed in Great Britain at the University Press, Cambridge

A catalogue record for this book is available from the British Library

Library of Congress cataloguing in publication data available

ISBN 0 521 43829 2 paperback

Cover illustration: Fjord in Southern Greenland © Simon Fraser, Science Photo Library

INTERGOVERNMENTAL PANEL ON CLIMATE CHANGE

CLIMATE CHANGE 1992
The Supplementary Report to the
IPCC Scientific Assessment

Report Prepared for IPCC by Working Group I

Edited by J.T.Houghton, B.A.Callander and S.K.Varney

(Meteorological Office, Bracknell, United Kingdom)

WMO

UNEP

Contents

Foreword

The Intergovernmental Panel on Climate Change (IPCC) was jointly established by our two organizations in 1988, under the Chairmanship of Professor Bert Bolin. The Panel formed three Working Groups:

(a) to assess the available scientific information on climate change,

(b) to assess the environmental and socio-economic impacts of climate change, and

(c) to formulate response strategies.

The Panel also established a Special Committee on the Participation of Developing Countries to promote the participation of those countries in its activities.

The IPCC First Assessment Report was completed in August 1990 and consists of the IPCC Scientific Assessment, the IPCC Impacts Assessment, the IPCC Response Strategies, the Policymakers' Summary of the IPCC Special Committee and the IPCC Overview. The Report has now become a standard work of reference, widely used by policymakers, scientists and other experts, and encompasses a remarkable coordinated effort by hundreds of specialists from all over the world.

Anticipating the need in 1992 for the latest information on climate change, in the context of the ongoing negotiations on the Framework Convention on Climate Change and the United Nations Conference on Environment and Development (Rio de Janeiro, June 1992), the Panel in 1991 requested its three Working Groups to produce updates to their 1990 Reports, addressing the key conclusions of the 1990 Assessment in the light of new data and analyses. The current publication is the result of the effort of Working Group I to produce an update of the IPCC Scientific Assessment.

As in 1990, success in producing this Report has depended upon the enthusiasm and cooperation of scientists worldwide. We admire, applaud, and are grateful for their commitment to the IPCC process. We also take this opportunity to express our gratitude to Professor Bolin for his very able leadership of the IPCC, and once again congratulate Sir John Houghton, Chairman of Working Group I, and his secretariat for another job well done.

G.O.P. Obasi
Secretary-General
World Meteorological Organization

M.K. Tolba
Executive Director
United Nations Environment Programme

Preface

The first Scientific Assessment of Climate Change prepared by Working Group I (WGI) of the IPCC[1] was published in June 1990[2]. That report was a comprehensive statement of the state of scientific knowledge concerning climate change and mankind's role therein, and resulted from almost two years' work by 170 scientists worldwide. A further 200 scientists were involved in its peer review.

In the eighteen months since then, scientific activity has continued to focus on the problem of climate change and significant progress has been made in a number of important areas. The purpose of this Supplement is to update the 1990 report, paying particular attention to its key conclusions and to new issues which have appeared in the scientific debate. The Supplement does not cover the full range of topics addressed by the 1990 report and should be read in conjunction with, rather than in place of, that report.

The conclusions presented in the Supplement are based entirely on the supporting scientific material published here, which has been prepared by leading scientists and exposed to a widespread and thorough peer review. To assist in the preparation of the report and the background papers a number of workshops were held, each devoted to one of the main topics being addressed. Generation of the background papers involved, either as lead authors or contributors, 118 scientists from 22 countries. A further

380 scientists from 63 countries and 18 UN or non-governmental organizations participated in the peer review of both the background material and the Supplement. The text of the Supplement was agreed in January 1992 at a plenary meeting of WGI held in Guangzhou, China, attended by 130 delegates from 47 countries. It can therefore be considered as an authoritative statement of the contemporary views of the international scientific community.

We are pleased to acknowledge the contributions of so many, in particular the Lead Authors who have given so freely of their expertise and time in the preparation of this report, together with the data and modelling centres who have provided so much information. We would also like to thank the core team at the Meteorological Office in Bracknell, which was responsible for organizing most of the workshops and for coordinating the activities of the very large number of contributing scientists. Members of the team and those who particularly contributed to its work were Bruce Callander, Shelagh Varney, Bob Watson, John Mitchell, Chris Folland, David Parker, Mavis Cook, Mavis Martin and Flo Ormond. Financial support for the team's work was provided by the United Kingdom's Department of the Environment.

Thanks are also due to all the members of the IPCC central secretariat at WMO, Geneva, under the direction of Dr. N. Sundararaman, for their friendly and tireless

---

[1] Organizational details of IPCC and Working Group I are shown in Appendix 1.

[2] *Climate Change. The IPCC Scientific Assessment.* J.T. Houghton, G.J. Jenkins and J.J. Ephraums (Eds.), Cambridge University Press, 1990, pp365.

assistance in workshop organization and coordination with the other IPCC working groups.

We are confident that this 1992 Supplement will assist further in building the firm scientific foundation necessary for the formulation of a rational and comprehensive response by mankind to the issue of climate change.

Sir John Houghton
Chairman, IPCC Working Group I

Meteorological Office
United Kingdom

Prof Bert Bolin
Chairman, IPCC

Stockholm University
Sweden

February 1992

The 1992 IPCC Supplement:
Scientific Assessment

Prepared by IPCC Working Group I

CONTENTS

1. Current Task

The fifth session of the Intergovernmental Panel on Climate Change (IPCC) (Geneva, March 1991) adopted six tasks for the ongoing work of its three working groups. While successful completion of these tasks required cooperation between all three groups, particular responsibility fell to the Scientific Assessment working group (WGI) for Tasks 1, 2 and 6:

Task 1: Assessment of net greenhouse gas emissions.
Sub-section 1: Sources and sinks of greenhouse gases.
Sub-section 2: Global Warming Potentials.

Task 2: Predictions of the regional distributions of climate change and associated impact studies; including model validation studies.

Task 6: Emissions scenarios.

The tasks were divided into long- and short-term components. The purpose of the short-term workplan, whose results are reported in the present document, was to provide an update to the 1990 IPCC Scientific Assessment, addressing some of the key issues of that report. This update is by definition less comprehensive than the 1990 report - for example sea level rise apart from the effect of thermal expansion is not included. It is against the background of that document that the findings of this update should be read.

This assessment, in order to incorporate as much recent material as possible, necessarily includes discussion of new results which have not yet been through, or are currently undergoing, the normal process of peer review. Where such is the case the provisional nature of the results has been taken into account.

A brief progress report on the preparation of guidelines for the compilation of national inventories of greenhouse gas emissions, part of WGI's long-term work under Task 1, appears as an Annex to this Supplement.

2. Our Major Conclusions

Findings of scientific research since 1990 do not affect our fundamental understanding of the science of the greenhouse effect and either confirm or do not justify alteration of the major conclusions of the first IPCC Scientific Assessment, in particular the following:

- emissions resulting from human activities are substantially increasing the atmospheric concentrations of the greenhouse gases: carbon dioxide, methane, chlorofluorocarbons, and nitrous oxide;
- the evidence from the modelling studies, from observations and the sensitivity analyses indicate that the sensitivity of global mean surface temperature to doubling CO_2 is unlikely to lie outside the range 1.5 to 4.5°C;

- there are many uncertainties in our predictions particularly with regard to the timing, magnitude and regional patterns of climate change;
- global mean surface air temperature has increased by 0.3 to 0.6°C over the last 100 years;
- the size of this warming is broadly consistent with predictions of climate models, but it is also of the same magnitude as natural climate variability. Thus the observed increase could be largely due to this natural variability; alternatively this variability and other human factors could have offset a still larger human-induced greenhouse warming;
- the unequivocal detection of the enhanced greenhouse effect from observations is not likely for a decade or more.

There are also a number of significant new findings and conclusions which we summarize as follows:

Gases and Aerosols

- Depletion of ozone in the lower stratosphere in the middle and high latitudes results in a decrease in radiative forcing which is believed to be comparable in magnitude to the radiative forcing contribution of chlorofluorocarbons (CFCs) (globally-averaged) over the last decade or so.
- The cooling effect of aerosols [†] resulting from sulphur emissions may have offset a significant part of the greenhouse warming in the Northern Hemisphere (NH) during the past several decades. Although this phenomenon was recognized in the 1990 report, some progress has been made in quantifying its effect.
- The Global Warming Potential (GWP) remains a useful concept but its practical utility for many gases depends on adequate quantification of the indirect effects as well as the direct. We now recognize that there is increased uncertainty in the calculation of GWPs, particularly in the indirect components and, whilst indirect GWPs are likely to be significant for some gases, the numerical estimates in this Supplementary Report are limited to direct GWPs.
- Whilst the rates of increase in the atmospheric concentrations of many greenhouse gases have continued to grow or remain steady, those of methane and some halogen compounds have slowed.

[†] The scientific definition of 'aerosol' is an airborne particle or collection of particles, but the word has become associated, erroneously, with the propellant used in 'aerosol sprays'. Throughout this report the term 'aerosol' means airborne particle or particles.

- Some data indicate that global emissions of methane from rice paddies may amount to less than previously estimated.

Scenarios

- Steps have been taken towards a more comprehensive analysis of the dependence of future greenhouse gas emissions on socio-economic assumptions and projections. A set of updated scenarios have been developed for use in modelling studies which describe a wide range of possible future emissions in the absence of a coordinated policy response to climate change.

Modelling

- Climate models have continued to improve in respect of both their physical realism and their ability to simulate present climate on large scales, and new techniques are being developed for the simulation of regional climate.
- Transient (time-dependent) simulations with coupled ocean-atmosphere models (CGCMs), in which neither aerosols nor ozone changes have been included, suggest a rate of global warming that is consistent, within the range of uncertainties, with the 0.3°C per decade warming rate quoted by IPCC (1990) for Scenario A of greenhouse gas emissions.
- The large-scale geographical patterns of warming produced by the transient model runs with CGCMs are generally similar to the patterns produced by the earlier equilibrium models except that the transient simulations show reduced warming over the northern North Atlantic and the southern oceans near Antarctica.
- CGCMs are capable of reproducing some features of atmospheric variability on intra-decadal time-scales.
- Our understanding of some climate feedbacks and their incorporation in the models has improved. In particular, there has been some clarification of the role of upper tropospheric water vapour. The role of other processes, in particular cloud effects, remains unresolved.

Climate Observations

- The anomalously high global mean surface temperatures of the late 1980s have continued into 1990 and 1991 which are the warmest years in the record.
- Average warming over parts of the Northern Hemisphere mid-latitude continents has been found to be largely characterized by increases in minimum (night-time) rather than maximum (daytime) temperatures.
- Radiosonde data indicate that the lower troposphere

has warmed over recent decades. Since meaningful trends cannot be assessed over periods as short as a decade, the widely reported disagreements between decadal trends of air temperature from satellite and surface data cannot be confirmed because the trends are statistically indistinguishable.

- The volcanic eruption of Mount Pinatubo in 1991 is expected to lead to transitory stratospheric warming. With less certainty, because of other natural influences, surface and tropospheric cooling may occur during the next few years.
- Average warming over the Northern Hemisphere during the last four decades has not been uniform, with marked seasonal and geographic variations; this warming has been especially slow, or absent, over the extratropical north west Atlantic.
- The consistency between observations of global temperature changes over the past century and model simulations of the warming due to greenhouse gases over the same period is improved if allowance is made for the increasing evidence of a cooling effect due to sulphate aerosols and stratospheric ozone depletion.

The above conclusions have implications for future projections of global warming and somewhat modify the estimated rate of warming of 0.3°C per decade for the greenhouse gas emissions Scenario A of the IPCC 1990 Report. If sulphur emissions continue to increase, this warming rate is likely to be reduced, significantly in the Northern Hemisphere, by an amount dependent on the future magnitude and regional distribution of the emissions. Because sulphate aerosols are very short-lived in the atmosphere, their effect on global warming rapidly adjusts to increases or decreases in emissions. It should also be noted that while partially offsetting the greenhouse warming, the sulphur emissions are also responsible for acid rain and other environmental effects. There is a further small net reduction likely in the rate of global warming during the next few decades due to decreases in stratospheric ozone, partially offset by increases in tropospheric ozone.

Research carried out since the 1990 IPCC Assessment has served to improve our appreciation of key uncertainties. There is a continuing need for increased monitoring and research into climate processes and modelling. This must involve, in particular, strengthened international collaboration through the World Climate Research Programme (WCRP), the International Geosphere Biosphere Programme (IGBP) and the Global Climate Observing System (GCOS).

HOW DOES THE CLIMATE SYSTEM WORK, AND WHAT INFORMATION DO WE NEED TO ESTIMATE FUTURE CHANGES?

How does the climate system work?

The Earth absorbs radiation from the Sun, mainly at the surface. This energy is then redistributed by the atmosphere and ocean and re-radiated to space at longer ('thermal', 'terrestrial' or 'infrared') wavelengths. Some of the thermal radiation is absorbed by radiatively-active ('greenhouse') gases in the atmosphere, principally water vapour, but also carbon dioxide, methane, the CFCs, ozone and other greenhouse gases. The absorbed energy is re-radiated in all directions, downwards as well as upwards such that the radiation that is eventually lost to space is from higher, colder levels in the atmosphere (see diagram below). The result is that the surface loses less heat to space than it would do in the absence of the greenhouse gases and consequently stays warmer than it would otherwise be. This phenomenon, which acts rather like a 'blanket' around the Earth, is known as the greenhouse effect.

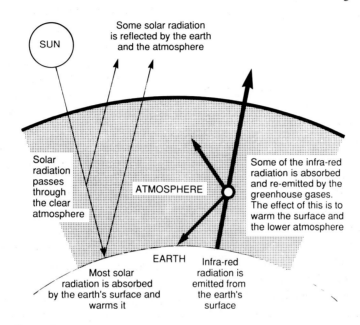

What factors can change climate?

Any factor which alters the radiation received from the Sun or lost to space, or which alters the redistribution of energy within the atmosphere, and between the atmosphere, land and ocean, will affect climate.

The Sun's output of energy is known to change by small amounts over an 11-year cycle, and variations over longer periods may occur. On time-scales of tens to thousands of years, slow variations in the Earth's orbit have led to changes in the seasonal and latitudinal distribution of solar radiation; these changes have played an important part in controlling the variations of past climate.

Increases in the concentration of the greenhouse gases will reduce the efficiency with which the Earth cools to space and will tend to warm the lower atmosphere and surface. The amount of warming depends on the size of the increase in concentration of each greenhouse gas, the radiative properties of the gases involved, and the concentration of other greenhouse gases already present in the atmosphere. It also can depend on local effects such as the variation with height of the concentration of the greenhouse gas, a consideration that may be particularly germane to water vapour which is not uniformly mixed throughout the atmosphere. The effect is not a simple one and the balance which is struck between these factors depends on many aspects of the climate system.

Aerosols (small particles) from volcanoes, emissions of sulphates from industry and other sources can absorb and reflect radiation. Moreover, changes in aerosol concentrations can alter cloud reflectivity through their effect on cloud properties. In most cases aerosols tend to cool climate. In general, they have a much shorter lifetime than greenhouse gases so their concentrations respond much more quickly to changes in emissions.

Any changes in the radiative balance of the Earth, including those due to an increase in greenhouse gases or in aerosols, will tend to alter atmospheric and oceanic temperatures and the associated circulation and weather patterns. However, climate varies naturally on all time-scales due to both external and internal factors. To distinguish man-made climate variations from those natural changes, it is necessary to identify the man-made 'signal' against the background 'noise' of natural climate variability.

A necessary starting point for the prediction of changes in climate due to increases in greenhouse gases and aerosols is an estimate of their future concentrations. This requires a knowledge of both the strengths of their sources (natural and man-made) and also the mechanisms of their eventual removal from the atmosphere (their sinks). The projections of future concentrations can then be used in climate models to estimate the climatic response. We also need to determine whether or not the predicted changes will be noticeable above the natural variations in climate. Finally, observations are essential in order to monitor climate, to study climatic processes and to help in the development and validation of models.

3.   How has our Understanding of the Sources and Sinks of Greenhouse Gases and Aerosols Changed?

During the last eighteen months there have been a number of important advances in our understanding of greenhouse gases and aerosols. These advances include an improved quantitative understanding of the atmospheric distributions, trends, sources and sinks of greenhouse gases, their precursors and aerosols, and an improved understanding of the processes controlling their global budgets.

Atmospheric Concentrations and Trends of Long-lived Greenhouse Gases:

The atmospheric concentrations of the major long-lived greenhouse gases [carbon dioxide (CO_2), methane (CH_4), nitrous oxide (N_2O), chlorofluorocarbons (CFCs), and carbon tetrachloride (CCl_4)] continue to increase because of human activities. While the growth rates of most of these gases have been steady or increasing over the past decade, that of CH_4 and some of the halocarbons has been decreasing. The rate for CH_4 has declined from about 20 ppbv/yr in the late 1970s to possibly as low as 10 ppbv/yr in 1989. While a number of hypotheses have been forwarded to explain these observations, none is completely satisfactory.

Atmospheric Concentrations and Trends of Other Gases that Influence the Radiative Budget:

Ozone (O_3) is an effective greenhouse gas both in the stratosphere and in the troposphere. Significant decreases have been observed during the last one to two decades in total column O_3 at all latitudes - except the tropics - in spring, summer and winter. The downward trends were larger during the 1980s than in the 1970s. These decreases have occurred predominantly in the lower stratosphere (below 25km), where the rate of decrease has been up to 10% per decade depending on altitude. In addition, there is evidence to indicate that O_3 levels in the troposphere up to 10km altitude above the few existing ozonesonde stations at northern middle latitudes have increased by about 10% per decade over the past two decades. Also, the abundance of carbon monoxide (CO) appears to be increasing in the NH at about 1% per year. However, there is little new information on the global trends of other tropospheric O_3 precursors, (non-methane hydrocarbons (NMHC) and oxides of nitrogen (NO_x)).

Sources and Sinks of Carbon Dioxide:

The two primary sources of the observed increase in atmospheric CO_2 are combustion of fossil fuels and land-use changes; cement production is a further important source.

The emission of CO_2 from the combustion of fossil fuels grew between 1987 and 1989. Preliminary data for 1990

indicate similar emissions to 1989. The best estimate for global fossil fuel emissions in 1989 and 1990 is 6.0±0.5 GtC [†], compared to 5.7±0.5 GtC in 1987 (IPCC, 1990). The estimated total release of carbon in the form of CO_2 from oil well fires in Kuwait during 1991 was 0.065 GtC, about 1% of total annual anthropogenic emissions.

The direct net flux of CO_2 from land-use changes (primarily deforestation), integrated over time, depends upon the area of land deforested, the rate of reforestation and afforestation, the carbon density of the original and replacement forests, and the fate of above-ground and soil carbon. These and other factors are needed to estimate annual net emissions but significant uncertainties exist in our quantitative knowledge of them. Since IPCC (1990) some progress has been made in reducing the uncertainties associated with the rate of deforestation, at least in Brazil. A comprehensive, multi-year, high spatial resolution satellite data set has been used to estimate that the average rate of deforestation in the Brazilian Amazonian forest between 1978 and 1989 was 2.1 million hectares (Mha) per year. The rate increased between 1978 and the mid-1980s, and has decreased to 1.4 Mha/yr in 1990. The Food and Agriculture Organization (FAO), using information supplied by individual countries, recently estimated that the rate of global tropical deforestation in closed and open canopy forests for the period 1981-1990 was about 17 Mha/yr, approximately 50% higher than in the period 1976-1980.

Despite the new information regarding rates of deforestation, the uncertainties in estimating CO_2 emissions are so large that there is no strong reason to revise the IPCC 1990 estimate of annual average net flux to the atmosphere of 1.6±1.0 GtC from land-use change during the decade of the 1980s.

Since IPCC (1990) particular attention has focussed on understanding the processes controlling the release and uptake of CO_2 from both the terrestrial biosphere and the oceans, and on the quantification of the fluxes. Based on models and the atmospheric distribution of CO_2, it appears that there is a small net addition of carbon to the atmosphere from the equatorial region, a combination of outgassing of CO_2 from warm tropical waters and a terrestrial biospheric component that is the residual between large sources (including deforestation) and sinks. There appears to be a strong Northern Hemisphere sink, containing both oceanic and terrestrial biospheric com-

---

[†] 1 GtC (gigatonne of carbon) equals one billion (one thousand million (10^9)) tonnes of carbon

ponents, and a weak Southern Hemisphere (SH) sink. The previous IPCC global estimate for an ocean sink of 2.0±0.8 GtC per year is still a reasonable one. The terrestrial biospheric processes which are suggested as contributing to the sinks are sequestration due to forest regeneration, and fertilization arising from the effects of both CO_2 and nitrogen (N), but none of these can be adequately quantified. This implies that the imbalance (of order 1-2 GtC/yr) between sources and sinks, i.e., "the missing sink", has not yet been resolved. This fact has significant consequences for estimates of future atmospheric CO_2 concentrations (see Section 5) and the analysis of the concept of the Global Warming Potential (see Section 6).

Sources of Methane:
A total (anthropogenic plus natural) annual emission of CH_4 of about 500Tg can be deduced from the magnitude of its sinks combined with its rate of accumulation in the atmosphere. While the sum of the individual sources is consistent with a total of 500Tg CH_4, there are still many uncertainties in accurately quantifying the magnitude of emissions from individual sources. Significant new information includes a revised rate of removal of CH_4 by atmospheric hydroxyl (OH) radicals (because of a lower rate constant), a new evaluation of some of the sources (e.g., from rice fields) and the addition of new sources (e.g., animal and domestic waste). Recent CH_4 isotopic studies suggest that approximately 100Tg CH_4 (20% of the total CH_4 source) is of fossil origin, largely from the coal, oil, and natural gas industries. Recent studies of CH_4 emissions from rice agriculture, in particular Japan, India, Australia, Thailand and China, show that the emissions depend on growing conditions, particularly soil characteristics, and vary significantly. While the overall uncertainty in the magnitude of global emissions from rice agriculture remains large, a detailed analysis now suggests significantly lower annual emissions than reported in IPCC (1990). The latest estimate of the atmospheric lifetime of CH_4 is about 11 years.

Sources of Nitrous Oxide:
Adipic acid (nylon) production, nitric acid production and automobiles with three-way catalysts have been identified as possibly significant anthropogenic global sources of nitrous oxide. However, the sum of all known anthropogenic and natural sources is still barely sufficient to balance the calculated atmospheric sink or to explain the observed increase in the atmospheric abundance of N_2O.

Sources of Halogenated Species:
The worldwide consumption of CFCs 11, 12, and 113 is now 40% below 1986 levels, substantially below the amounts permitted under the Montreal Protocol. Further reductions are mandated by the 1990 London Amendments to the Montreal Protocol. As CFCs are phased out, HCFCs and HFCs will substitute, but at lower emission rates.

Stratospheric Ozone Depletion:
Even if the control measures of the 1990 London Amendments to the Montreal Protocol were to be implemented by all nations, the abundance of stratospheric chlorine and bromine will increase over the next several years. The Antarctic ozone hole, caused by industrial halocarbons, will therefore recur each spring. In addition, as the weight of evidence suggests that these gases are also responsible for the observed reductions in middle and high latitude stratospheric O_3, the depletion at these latitudes is predicted to continue unabated through the 1990s.

Sources of Precursors of Tropospheric Ozone:
Little new information is available regarding the tropospheric ozone precursors (CO, NMHC, and NO_x), all of which have significant natural and anthropogenic sources. Their detailed budgets therefore remain uncertain.

Source of Aerosols:
Industrial activity, biomass burning, volcanic eruptions, and sub-sonic aircraft contribute substantially to the formation of tropospheric and stratospheric aerosols. Industrial activities are concentrated in the Northern Hemisphere where their impact on tropospheric sulphate aerosols is greatest. Sulphur emissions, which are due in large part to combustion effluents, have a similar emissions history to that of anthropogenic CO_2. Estimates of emissions of natural sulphur compounds have been reduced from previous figures, thereby placing more emphasis on the anthropogenic contribution.

4. Scenarios of Future Emissions
Scenarios of net greenhouse gas and aerosol precursor emissions for the next 100 years or more are necessary to support study of potential anthropogenic impacts on the climate system. The scenarios provide inputs to climate models and assist in the examination of the relative importance of relevant trace gases and aerosol precursors in changing atmospheric composition and climate. Scenarios can also help in improving the understanding of key relationships among factors that drive future emissions.

Scenario outputs are not predictions of the future, and should not be used as such; they illustrate the effect of a wide range of economic, demographic and policy assumptions. They are inherently controversial because they reflect different views of the future. The results of scenarios can vary considerably from actual outcomes even over short time horizons. Confidence in scenario outputs decreases as the time horizon increases, because the basis for the underlying assumptions becomes increasingly speculative. Considerable uncertainties surround the evolution of the types and levels of human activities (including economic growth and structure), technological advances, and human responses to possible environmental, economic and institutional constraints. Consequently, emission scenarios must be constructed carefully and used with great caution.

Since completion of the 1990 IPCC Scenario A (SA90) events and new information have emerged which relate to that scenario's underlying assumptions. These developments include: the London Amendments to the Montreal Protocol; revision of population forecasts by the World Bank and United Nations; publication of the IPCC Energy and Industry Sub-group scenario of greenhouse gas emissions to 2025; political events and economic changes in the former USSR, Eastern Europe and the Middle East; re-estimation of sources and sinks of greenhouse gases (reviewed in this Assessment); revision of preliminary FAO data on tropical deforestation; and new scientific studies on forest biomass. There has also been recognition of considerable uncertainty regarding other important factors that drive future emissions.

These factors have led to an update of the SA90. Six alternative IPCC Scenarios (IS92a-f) now embody a wide array of assumptions, summarized in Table 1, affecting how future greenhouse gas emissions might evolve in the absence of climate policies beyond those already adopted. This constitutes a significant improvement over the previous methodology. However, the probability of any of the resulting emission paths has not been analysed. IPCC WGI does not prefer any individual scenario. Other combinations of assumptions could illustrate a broader variety of emission trajectories. The different worlds which the new scenarios imply, in terms of economic, social and environmental conditions, vary widely. The current exercise provides an interim view and lays a basis for a more complete study of future emissions of greenhouse gas and aerosol precursors.

Scenario Results:

The range of possible greenhouse gas futures is very wide, as Figure 1 on page 13 illustrates (showing only CO_2). All six scenarios can be compared to SA90. IS92a is slightly lower than SA90 due to modest and largely offsetting changes in the underlying assumptions. (For example, compared to SA90, higher population forecasts increase the emission estimates, while phaseout of halocarbons and more optimistic renewable energy costs reduce them.) The highest greenhouse gas levels result from the new scenario IS92e which combines, among other assumptions, moderate population growth, high economic growth, high

fossil fuel availability and eventual hypothetical phaseout of nuclear power. The lowest greenhouse gas levels result from IS92c which assumes that population grows, then declines by the middle of the next century, that economic growth is low and that there are severe constraints on fossil fuel supplies. The results of all six scenarios appear in Table 2. Overall, the scenarios indicate that greenhouse gas emissions could rise substantially over the coming century in the absence of new measures explicitly intended to reduce their emission. However, IS92c has a CO_2 emission path which eventually falls below its starting 1990 level. IS92b, a modification of IS92a, suggests that current commitments by many OECD Member countries to stabilize or reduce CO_2 might have a small impact on greenhouse gas emissions over the next few decades, but would not offset substantial growth in possible emissions in the long run. IS92b does not take into account that such commitments could accelerate development and diffusion of low greenhouse gas technologies, nor possible resulting shifts in industrial mix.

Carbon Dioxide:

The new emissions scenarios for CO_2 from the energy sector span a broad range of futures (see Figure 1).

Population and economic growth, structural changes in economies, energy prices, technological advance, fossil fuel supplies, nuclear and renewable energy availability are among the factors which could exert major influence on future levels of CO_2 emissions. Developments such as those in the republics of the former Soviet Union and in Eastern Europe, now incorporated into all the scenarios, have important implications for future fossil fuel carbon emissions, by affecting the levels of economic activities and the efficiency of energy production and use. Biotic carbon emissions in the early decades of the scenarios are higher than SA90, reflecting higher preliminary FAO estimates of current rates of deforestation in many - though not all - parts of the world, and higher estimates of forest biomass.

Halocarbons:

The revised scenarios for CFCs and other substances which deplete stratospheric ozone are much lower than in SA90. This is consistent with wide participation in the controls under the 1990 London Amendments to the Montreal Protocol. However, the future production and composition of CFC substitutes (HCFCs and HFCs) could significantly affect the levels of radiative forcing from these compounds.

Methane, Nitrous Oxide, Ozone Precursors and Sulphur Gases:

The distribution of CH_4 and N_2O emissions from the different sources has changed from the SA90 case.

Table 1: *Summary of Assumptions in the Six IPCC 1992 Alternative Scenarios.* [†]

Scenario	Population	Economic Growth	Energy Supplies [††]	Other [†††]	CFCs
IS92a	World Bank 1991 11.3 B by 2100	1990-2025: 2.9% 1990-2100: 2.3%	12,000 EJ Conventional Oil 13,000 EJ Natural Gas Solar costs fall to $0.075/kWh 191 EJ of biofuels available at $70/barrel	Legally enacted and internationally agreed controls on SOx, NOx and NMVOC emissions.	Partial compliance with Montreal Protocol. Technological transfer results in gradual phase out of CFCs also in non-signatory countries by 2075.
IS92b	World Bank 1991 11.3 B by 2100	1990-2025 2.9% 1990-2100 2.3%	Same as "a"	Same as "a" plus commitments by many OECD countries to stabilize or reduce CO_2 emissions.	Global compliance with scheduled phase out of Montreal Protocol.
IS92c	UN Medium Low Case 6.4 B by 2100	1990-2025 2.0% 1990-2100 1.2%	8,000 EJ Conventional Oil 7,300 EJ Natural Gas Nuclear costs decline by 0.4% annually.	Same as "a"	Same as "a"
IS92d	UN Medium Low Case 6.4 B by 2100	1990-2025 2.7% 1990-2100 2.0%	Oil and gas same as "c" Solar costs fall to $0.065/kWh 272 EJ of biofuels available at $50/barrel	Emission controls extended worldwide for CO, NO_x, NMVOC, and SO_x. Halt deforestation. Capture and use of emissions from coal mining and gas production and use.	CFC production phase out by 1997 for industrialized countries. Phase out of HCFCs.
IS92e	World Bank 1991 11.3 B by 2100	1990-2025 3.5% 1990-2100 3.0%	18,400 EJ conventional oil Gas same as "a" Phase out nuclear by 2075	Emission controls (30 % pollution surcharge on fossil energy).	Same as "d"
IS92f	UN Medium High Case 17.6 B by 2100	Same as "a"	Oil and gas same as "e" Solar costs fall to $0.083/kWh Nuclear costs increase to $0.09/kWh	Same as "a"	Same as "a"

[†] The assumptions for the 1990 Scenario A are described in IPCC (1990) Annex A, pp.331-339 .

[††] All scenarios assume coal resources up to 197,000 EJ. Up to 15% of this resource is assumed to be available at $1.30/gigajoule at the mine.

[†††] Tropical deforestation rates (for closed and open forests) begin from an average rate of 17.0 million hectares/year (FAO, 1991) for 1981-1990, then increase with population until constrained by availability of land not legally protected. IS91d assumes an eventual halt of deforestation for reasons other than climate. Above-ground carbon density per hectare varies with forest type from 16 to 117 tons C/hectare, with soil C ranging from 68 to 100 tons C/ha. However, only a portion of carbon is released over time with land conversion, depending on type of land conversion.

Table 2: Selected Results of Six 1992 IPCC Greenhouse Gas Scenarios

Scenario	Years	Decline in TPER/GNP (average annual change)	Decline in C intensity (average annual change)	Cumulative Net Fossil C Emissions (GtC)	Tropical Deforestation		Year	Emissions Per Year				
					Total Forest Cleared (million hectares)	Cumulative Net C Emissions (GtC)		CO$_2$ (GtC)	CH$_4$ (Tg)	N$_2$O (Tg N)	CFCs (kt)	SO$_x$ (Tg S)
IS92a	1990-2025	0.8%	0.4%	285	678	42	1990	7.4	506	12.9	827	98
	1990-2100	1.0%	0.2%	1386	1447	77	2025	12.2	659	15.8	217	141
							2100	20.3	917	17.0	3	169
IS92b	1990-2025	0.9%	0.4%	275	678	42	2025	11.8	659	15.7	36	140
	1990-2100	1.0%	0.2%	1316	1447	77	2100	19.1	917	16.9	0	164
IS92c	1990-2025	0.6%	0.7%	228	675	42	2025	8.8	589	15.0	217	115
	1990-2100	0.7%	0.6%	672	1343	70	2100	4.6	546	13.7	3	77
IS92d	1990-2025	0.8%	0.9%	249	420	25	2025	9.3	584	15.1	24	104
	1990-2100	0.8%	0.7%	908	651	30	2100	10.3	567	14.5	0	87
IS92e	1990-2025	1.0%	0.2%	330	678	42	2025	15.1	692	16.3	24	163
	1990-2100	1.1%	0.2%	2050	1447	77	2100	35.8	1072	19.1	0	254
IS92f	1990-2025	0.8%	0.1%	311	725	46	2025	14.4	697	16.2	217	151
	1990-2100	1.0%	0.1%	1690	1686	93	2100	26.6	1168	19.0	3	204

TPER = Total Primary Energy Requirement
Carbon intensity is defined as units of carbon per unit of TPER
CFCs include CFC-11, CFC-12, CFC-113, CFC-114 and CFC-115

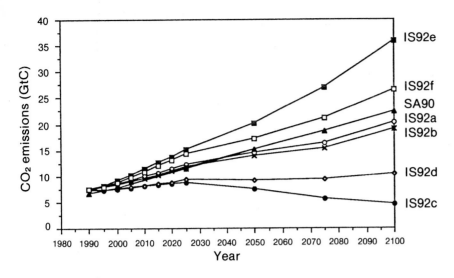

Figure 1: CO_2 emissions from energy, cement production and deforestation.

Methane emissions from rice paddies are lower, and emissions from animal waste and domestic sewage have been added. N_2O emission factors for stationary sources and biomass burning have been revised downwards. Adipic and nitric acid have been included as additional sources of N_2O. Preliminary analysis of the emissions of volatile organic compounds and sulphur dioxide suggests that the global emissions of these substances are likely to grow in the coming century if no new limitation strategies are implemented.

5. Relationship Between Emissions and Atmospheric Concentrations and the Influence on the Radiative Budget

A key issue is to relate emissions of greenhouse gases, greenhouse gas precursors and aerosol precursors to future concentrations of greenhouse gases and aerosols in order to assess their impact on the radiative balance. A number of different types of model have been developed.

Carbon Cycle Models:

While there is a variety of carbon cycle models (including 3-D ocean-atmosphere models, l-D ocean-atmosphere box-diffusion models, and box models that incorporate a terrestrial biospheric sink) all such models are subject to considerable uncertainty because of an inadequate understanding of the processes controlling the uptake and release of CO_2 from the oceans and terrestrial ecosystems. Some models assume a net neutral terrestrial biosphere, balancing fossil fuel emissions of CO_2 by oceanic uptake and atmospheric accumulation, others achieve balance by invoking additional assumptions regarding the effect of CO_2 fertilization on the different parts of the biosphere. However, even models that balance the past and

contemporary carbon cycle may not predict future atmospheric concentrations accurately because they do not necessarily represent the proper mix of processes on land and in the oceans. The differences in predicted changes in CO_2 concentrations are up to 30%. This does not represent the major uncertainty in the prediction of future climate change compared with uncertainties in estimating future patterns of trace gas emissions, and in quantifying climate feedback processes. A simple empirical estimate can be based on the assumption that the fraction of emissions which remains in the atmosphere is the same as that observed over the last decade; i.e. 46±7%.

Atmospheric Gas Phase Chemistry Models:

Current tropospheric models exhibit substantial differences in their predictions of changes in O_3, in the hydroxyl radical (OH) and in other chemically active gases due to emissions of CH_4, non-methane hydrocarbons, CO and, in particular, NO_x. These arise from uncertainties in the knowledge of background chemical composition and from our inability to represent small-scale processes occurring within the atmosphere. These deficiencies limit the accuracy of predicted changes in the abundance and distribution of tropospheric O_3, and in the lifetimes of a number of other greenhouse gases, including the HCFCs and HFCs, all of which depend upon the abundance of the OH radical. Increases in CH_4, NMHCs, and CO all lead to increases in O_3, and decreases in OH, thus leading to an increase in radiative forcing. On the other hand, because increases in NO_x lead to an increase in both O_3 and OH, the net effect of NO_x on radiative forcing is uncertain.

Atmospheric Sulphate Aerosol Models:

The atmospheric chemistry of sulphate aerosols and their precursors has been extensively studied in relation to the acid rain issue. While our understanding of processes related to chemical transformations has increased significantly in recent years, substantial uncertainties remain, especially regarding the microphysics of aerosol formation, interaction of aerosols with clouds, and the removal of aerosol particles by precipitation.

6. How has our Understanding of Changes in Radiative Forcing Changed?

Since IPCC (1990), there have been significant advances in our understanding of the impact of ozone depletion and sulphate aerosols on radiative forcing and of the limitations of the concept of the Global Warming Potential.

Radiative Forcing due to Changes in Stratospheric Ozone:

For the first time observed global depletions of O_3 in the lower stratosphere have been used to calculate changes in the radiative balance of the atmosphere. Although the results are sensitive to atmospheric adjustments, and no GCM studies of the implications of the O_3 changes on surface temperature have been performed, the radiative balance calculations indicate that the O_3 reductions observed during the 1980s have caused reductions in the radiative forcing of the surface-troposphere system at middle and high latitudes. This reduction in radiative forcing resulting from O_3 depletion could, averaged on a global scale and over the last decade, be approximately equal in magnitude and opposite in sign to the enhanced radiative forcing due to increased CFCs during the same time period. The effect at high latitudes is particularly pronounced and, because of these large variations with latitude and region, studies using GCMs are urgently required to further test these findings.

Radiative Forcing due to Changes in Tropospheric Ozone:

While there are consistent observations of an increase in tropospheric ozone (up to 10% per decade) at a limited number of locations in Europe, there is not an adequate global set of observations to quantify the magnitude of the increase in radiative forcing. However, it has been calculated that a 10% uniform global increase in tropospheric ozone would increase radiative forcing by about a tenth of a watt per square metre.

Radiative Effects of Sulphur Emissions:

Emissions of sulphur compounds from anthropogenic sources lead to the presence of sulphate aerosols which reflect solar radiation. This is likely to have a cooling influence on the Northern Hemisphere (there is negligible effect in the Southern Hemisphere). For clear-sky conditions alone, the cooling caused by current rates of emissions has been estimated to be about $1Wm^{-2}$ averaged over the Northern Hemisphere, a value which should be compared with the estimate of $2.5Wm^{-2}$ for the heating due to anthropogenic greenhouse gas emissions up to the present. The non-uniform distribution of anthropogenic sulphate aerosols coupled with their relatively short atmospheric residence time produce large regional variations in their effects. In addition, sulphate aerosols may affect the radiation budget through changes in cloud optical properties.

Global Warming Potentials:

Gases can exert a radiative forcing both directly and indirectly: direct forcing occurs when the gas itself is a greenhouse gas; indirect forcing occurs when chemical transformation of the original gas produces a gas or gases which themselves are greenhouse gases. The concept of the Global Warming Potential (GWP) has been developed for policymakers as a measure of the possible warming effect on the surface-troposphere system arising from the emission of each gas relative to CO_2. The indices are calculated for the contemporary atmosphere and do not take into account possible changes in chemical composition of the atmosphere. Changes in radiative forcing due to CO_2 are non-linear with changes in the atmospheric CO_2 concentrations. Hence, as CO_2 levels increase from present values, the GWPs of the non-CO_2 gases would be higher than those evaluated here. For the concept to be most useful, both the direct and indirect components of the GWP need to be quantified.

Direct Global Warming Potentials:

The direct components of the Global Warming Potentials (GWPs) have been recalculated, taking into account revised estimated lifetimes, for a set of time horizons ranging from 20 to 500 years, with CO_2 as the reference gas. The same ocean-atmosphere carbon cycle model as in IPCC (1990) has been used to relate CO_2 emissions to concentrations. Table 3 shows values for a selected set of key gases for the 100 year time horizon. While in most cases the values are similar to the previous IPCC (1990) values, the GWPs for some of the HCFCs and HFCs have increased by 20 to 50% because of revised estimates of their lifetimes. The direct GWP of CH_4 has been adjusted upward, correcting an error in the previous IPCC report. The carbon cycle model used in these calculations probably underestimates both the direct and indirect GWP values for all non-CO_2 gases. The magnitude of the bias depends on the atmospheric lifetime of the gas, and the GWP time horizon.

Indirect Global Warming Potentials:

Because of our incomplete understanding of chemical processes, most of the indirect GWPs reported in IPCC (1990) are likely to be in substantial error, and none of

Table 3: *Direct GWPs for 100-year Time Horizon*

Gas	Direct Global Warming Potential (GWP)	Sign of the Indirect Component of the GWP
Carbon dioxide	1	none
Methane	11	positive
Nitrous oxide	270	uncertain
CFC-11	3400	negative
CFC-12	7100	negative
HCFC-22	1600	negative
HFC-134a	1200	none

them can be recommended. Although we are not yet in a position to recommend revised numerical values, we know, however, that the indirect GWP for methane is positive and could be comparable in magnitude to its direct value. In contrast, based on the sub-section above, the indirect GWPs for chlorine and bromine halocarbons are likely to be negative. The concept of a GWP for short-lived, inhomogeneously distributed constituents, such as CO, NMHC, and NO_x may prove inapplicable, although, as noted above, we know that these constituents will affect the radiative balance of the atmosphere through changes in tropospheric ozone and OH. Similarly, a GWP for SO_2 is viewed to be inapplicable because of the non-uniform distribution of sulphate aerosols.

Influence of Changes in Solar Output:

The existence of strong correlations between characteristics of the solar activity cycle and global mean temperature has been reported. The only immediately plausible physical explanation of these correlations involves variability of the sun's total irradiance on time-scales longer than that of the 11-year activity cycle. Since precise measurements of the irradiance are only available for the last decade, no firm conclusions regarding the influence of solar variability on climate change can be drawn.

7. Confidence in Model Predictions

There continues to be slow improvement in the ability of models to simulate present climate, although further improvements in the model resolution and the parametrization of physical processes are needed. Since the last report, further evidence has accumulated that atmospheric models are capable of reproducing a range of aspects of atmospheric variability. Coupled ocean-atmosphere models produce variability on decadal time-scales similar in some respects to that observed, and ocean models show longer term fluctuations associated with changes in the thermohaline circulation.

There has been some clarification of the nature of water vapour feedback, although the radiative effects of clouds and related processes continue to be the major source of uncertainty and there remain uncertainties in the predicted changes in upper tropospheric water vapour in the tropics. Biological feedbacks have not yet been taken into account in simulations of climate change.

Increased confidence in the geographical patterns of climate change will require new simulations with improved coupled models and with radiative forcing scenarios that include aerosols.

Confidence in regional climate patterns based directly on GCM output remains low and there is no consistent evidence regarding changes in variability or storminess. GCM results can be interpolated to smaller scales using statistical methods (correlating regional climate with the large-scale flow) or a nested approach (high-resolution, regional climate models driven by large-scale GCM results). Both methods show promise but an insufficient number of studies have yet been completed to give an improved global picture of regional climate change due to increases in greenhouse gases; in any event both interpolation methods depend critically on the quality of the large-scale flow in the GCM. Given our incomplete knowledge of climate, we cannot rule out the possibility of surprises.

8. Simulated Rates of Change in Climate and their Geographical Distribution

Results of General Circulation Models (GCMs) available to the 1990 report mainly concerned *equilibrium* simulations. Only one *transient* model run (i.e., where the time-varying response of the climate to steadily increasing greenhouse gas concentrations is simulated) had been completed.

Since then many papers have appeared in the refereed literature concerned with climate models and their results. Significant progress has been made in the area of transient models, with four modelling groups having carried out

WHAT TOOLS DO WE USE AND WHAT INFORMATION DO WE NEED TO PREDICT FUTURE CLIMATE?

Models

The most highly developed tool which we have to model climate and climate change is known as a **general circulation model or GCM**. These models are based on the laws of physics and use descriptions in simplified physical terms (called parametrizations) of the smaller-scale processes such as those due to clouds and deep mixing in the ocean. 'Coupled' general circulation models (CGCMs) have the atmospheric component linked to an oceanic component of comparable complexity.

Climate forecasts are derived in a different way from weather forecasts. A weather prediction model gives a description of the atmosphere's state up to 10 days or so ahead, starting from a detailed description of an initial state of the atmosphere at a given time. Such forecasts describe the movement and development of large weather systems, though they cannot represent very small-scale phenomena; for example, individual shower clouds.

To estimate the influence of greenhouse gases or aerosols in changing climate, the model is first run for a few (simulated) decades. The statistics of the model's output are a description of the model's simulated climate which, if the model is a good one and includes all the important forcing factors, will bear a close resemblance to the climate of the real atmosphere and ocean. The above exercise is then repeated with increasing concentrations of the greenhouse gases or aerosols in the model. The differences between the statistics of the two simulations (for example in mean temperature and interannual variability) provide an estimate of the accompanying climate change.

We also need to determine whether or not the predicted changes will be noticeable above the natural variations in climate. Finally, observations are required in order to monitor climate, to improve the understanding of climate processes and to help in the validation of models.

The long-term change in surface air temperature following a doubling of carbon dioxide (referred to as the **climate sensitivity**) is generally used as a benchmark to compare models. The range of values for climate sensitivity reported in the 1990 Assessment, and re-affirmed in this Supplement, was 1.5 to 4.5°C, with a best estimate, based on model results and taking into account the observed climate record, of 2.5°C.

Simpler models, which simulate the behaviour of GCMs, are also used to make predictions of the evolution with time of global temperature from a number of emission scenarios. These so-called box-diffusion models contain highly simplified physics but give similar results to GCMs when globally averaged. Only comprehensive GCMs, however, can provide three-dimensional distributions of the changes in other climate variables, including the changes due to non-linear processes that are not given by simplified models. The extraction of this information from the results of coupled GCMs has only just begun.

Future concentrations of greenhouse gases and aerosols

A necessary starting point for the prediction of changes in climate due to changes in atmospheric constituents is an estimate of their future concentrations. This requires a knowledge of their sources and sinks (natural and man-made) and an estimate of how the strengths of these sources and sinks might change in the future (an **emissions scenario**). The projections of future concentrations can then be used in climate models to estimate the climatic response.

Do GCMs predict future climate?

To make a *prediction* of future climate it is necessary to fulfil two conditions: (a) include all of the major human and natural factors known to affect climate, and (b) predict the future magnitudes of atmospheric concentrations of greenhouse gases. So far, GCMs (and CGCMs) have included only radiative forcing induced by greenhouse gases, and therefore their results relate only to the greenhouse gas component of climate change.

At the time of the 1990 IPCC Report it was recognized that sulphate aerosols exert a significant negative radiative forcing on climate but this forcing was not well quantified. Since then progress has been made in understanding radiative forcing by sulphate aerosols, and an additional source of negative forcing has been identified in the depletion of stratospheric ozone due to halocarbons. The lack of these negative forcing factors in GCMs does not negate the results obtained from them so far. For example the estimates of climate sensitivity, which is defined purely in terms of CO_2 concentrations, are unchanged, and it is still believed that anthropogenic greenhouse gases, now and even more so in the future, represent the largest perturbation to the natural radiative balance of the atmosphere. However, it does mean that the rates of change of, say, surface temperature do need to be adjusted for additional forcing factors before they can fulfil condition (a). The second condition is fulfilled when we use a specific prediction (as opposed to a scenario) of future atmospheric concentrations of greenhouse gases.

climate simulations for up to 100 years using coupled atmosphere-ocean global climate models (CGCMs) which incorporate a detailed description of the deep ocean and therefore can simulate the climate lag induced by the deep ocean circulation. These models require substantial adjustments to fluxes of heat and fresh water in order to achieve a realistic simulation of present climate and this may distort the models' response to small perturbations such as those associated with increasing greenhouse gases. For simulations of future climate with these models, carbon dioxide concentrations have been increased at rates close to 1% per year (approximately the equivalent radiatively to the current rate of increase of greenhouse gases).

Internal variability obscures the geographical patterns of change during the first few decades of the experiments. However, once these patterns become established, they vary relatively little as the integrations progress and are similar to those produced by equilibrium models in a number of ways, for instance:

(i)   surface air temperatures increase more over land than over oceans;

(ii) precipitation increases on average at high latitudes, in the monsoon region of Asia and in the winter at mid-latitudes;

(iii) over some mid-latitude continental areas values of soil moisture are lower on average in summer.

The transient CO_2 simulations, however, show that over the northern North Atlantic and the southern oceans near Antarctica the warming is reduced by 60% or more relative to equilibrium simulations at the time of doubling CO_2.

Much further development and validation of coupled models is required.

9. What Would We Now Estimate for Climate Change?

The new simulations using coupled ocean-atmosphere GCMs, which do not include the effects of sulphates and ozone depletion, generally confirm the IPCC (1990) estimates of future warming at rates of about 0.3°C/decade (range 0.2 to 0.5°C/decade) over the next century for IPCC 1990 Scenario A. Because GCMs do not yet include possible opposing anthropogenic influences, including the forcing from sulphate aerosols and stratospheric ozone depletion, the net rate of increase in surface temperature is expected to be less, at least during the period for which sulphur emissions continue to increase, than would be expected from greenhouse gas forcing alone. However, the globally averaged magnitude of the effect of sulphate aerosols has not yet been calculated accurately and further work is needed.

The simulated rate of change of sea level *due to oceanic thermal expansion only* ranges from 2 to 4cm/decade, again consistent with the previous report.

New IPCC 1992 emissions scenarios (IS92a-f; see Section 4) have been derived in the light of new information and international agreements. In order to provide an initial assessment of the effect of the new scenarios, the change in surface temperature has been estimated with the simple climate model used in the IPCC 1990 report which has been calibrated against the more comprehensive coupled ocean-atmosphere models (see page 16 for a description of models). These calculations include, in the same way as did the 1990 calculations, the direct radiative forcing effects of all the greenhouse gases included in the scenarios. The effect of stratospheric ozone depletion and of sulphate aerosols have not been included, which again parallels the 1990 calculations. Figure 2 (see page 18) shows (a) the temporal evolution of surface temperature for IS92a, assuming the high, "best estimate" and low climate sensitivities (4.5, 2.5 and 1.5°C), and (b) the temperature changes for the six 1992 IPCC scenarios and the 1990 Scenario A, assuming the "best estimate" of climate sensitivity (see page 16 for the definition of climate sensitivity).

10. The Updated Record of Global Mean Temperatures

Continuing research into the nineteenth century ocean temperature record has not significantly altered our calculation of surface temperature warming over the past 100-130 years of 0.45±0.15°C. Furthermore, global surface temperatures for 1990 and 1991 have been similar to those of the warmest years of the 1980s and continue to be warmer than the rest of the record. The research has, however, led to a small adjustment in hemispheric temperatures. The long-term warming trends assessed in each hemisphere are now more nearly equal, with the Southern Hemisphere marginally warmer in the late nineteenth century and the Northern Hemisphere trend unchanged from previous estimates.

A notable feature over considerable areas of the continental land masses of the Northern Hemisphere is that warming over the last few decades is primarily due to an increase of night-time rather than daytime temperatures. These changes appear to be partly related to increases in cloudiness but other factors cannot be excluded such as a direct cooling effect of aerosols on maximum temperatures in sunny weather, an influence of increasing concentrations of greenhouse gases and some residual influence of urbanization on minimum temperatures. A more complete study is needed as only 25% of the global land area has been analysed. In this regard, regional changes in maximum, minimum and mean temperature related to changes in land use (e.g., desertification, deforestation or widespread irrigation) may need to be identified separately.

(a)

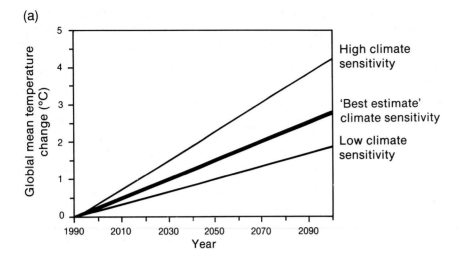

(b)

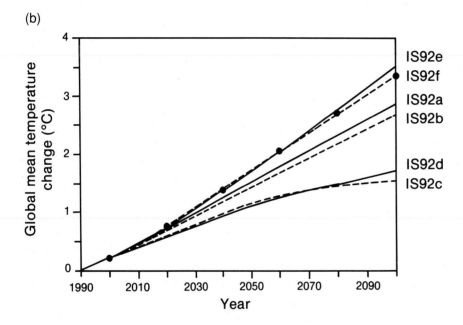

Figure 2: (a) Temperature change under scenario IS92a, (b) Best estimate temperature changes under IS92 and SA90. Solid circles show SA90.

A new source of information that supports higher sea surface temperatures in many tropical regions over the last decade concerns evidence that the bleaching of tropical corals has apparently increased. Bleaching has been shown to be related (in part) to episodes of sea surface temperature warmer than the normal range of tolerance of these animals, though increasing pollution may be having an influence.

There has been considerable interest in mid-tropospheric temperature observations made since 1979 from the Microwave Sounding Unit (MSU) aboard the TIROS-N satellites. The MSU data have a truly global coverage but there is only a short record (13 years) of measurements; the surface and the radiosonde data are less spatially complete but have much longer records (over 130 and near 30 years respectively). Globally-averaged trends in MSU, radiosonde and surface data sets between 1979 and 1991 differ somewhat (0.06, 0.17, and 0.18°C/decade, respectively), although the differences are not statistically significant. Satellite sounders, radiosonde and surface instruments all have different measurement characteristics and, in addition, geographical and temporal variations in mid-tropospheric and surface temperatures are not expected to be identical. Despite this, correlations between global annual values of the three data sets are quite high.

Note that it is not possible to rank recent warm years in an absolute way; it depends on which record is used, what level is referred to and how much uncertainty is attached to each value.

MSU data have been able to detect the impact on lower stratospheric temperature of volcanic eruptions in a striking way. Variability in these data between 1979 and 1991 is dominated by short-term temperature fluctuations (greatest in the tropics) following the injection of large amounts of aerosol into the stratosphere by the eruptions of El Chichon (1982) and Mt. Pinatubo (1991). Globally, temperature rises in the lower stratosphere were about 1°C and 1.3°C respectively; stratospheric warming due to El Chichon lasted nearly two years while that due to Mt. Pinatubo is still underway. The longer radiosonde record, however, shows a significant global cooling trend of about 0.4°C per decade since the middle 1960s in the lower stratosphere.

11. Are There any Trends in Other Climatically Important Quantities?

Precipitation variations of practical significance have been documented on many time and space scales, but due to data coverage and inhomogeneity problems nothing can be said about global-scale changes. An apparent increase of water vapour in the tropics parallels the increase in lower tropospheric temperature but it is not yet possible to say to what extent the changes are real and whether they are larger than natural variability.

A small, irregular, decrease of about 8% has been observed in annually averaged snow cover extent over the Northern Hemisphere since 1973 in a new, improved, compilation of these data. The reduction is thought to be real because annual values of snow cover extent and surface air temperatures over the extratropical Northern Hemisphere land have a high correlation of -0.76.

There is evidence that, regionally, relatively fast (sometimes called abrupt) climate changes can occur. These changes may persist for up to several decades but are often a function of season. These fast changes are poorly understood, but can be of considerable practical importance.

12. Are the Observed Temperature Changes Consistent with Predicted Temperature Changes?

CGCMs, which do not yet take into account changes in aerosols, predict a greater degree of warming in the Northern Hemisphere (NH) than in the Southern Hemisphere (SH), a result of the greater extent of land in the NH which responds more rapidly to forcing. The observed larger warming of the SH in recent decades (0.3°C between 1955 and 1985) than in the NH (which hardly warmed at all over the same period) is at first sight in conflict with this prediction. Recently, however, the NH has started to warm quite rapidly. The reasons for the differences in observed warming rates in the two hemispheres are not known though man-made aerosols (see Section 6) and changes in ocean circulation may have played a part.

Furthermore, increases in CFCs may have reduced ozone levels sufficiently to offset in a globally-averaged sense the direct greenhouse effect of CFCs. Consequently, the estimates of warming over the last 100 years due to increases in greenhouse gases made in the original report may be somewhat too rapid because they did not take account of these cooling influences. Taking this into account could bring the results of model simulations closer to the observed changes.

Individual volcanic eruptions, such as that of El Chichon, may have led to surface cooling over several years but should have negligible effect on the long-term trend. Some influence of solar variations on time-scales associated with several sunspot cycles remains unproven but is a possibility.

The conclusion of the 1990 report remains unchanged:
"the size of this warming is broadly consistent with predictions of climate models, but it is also of the same magnitude as natural climate variability. Thus the observed increase could be largely due to this natural variability; alternatively this variability and other human factors could have offset a still larger human-induced greenhouse warming".

13. Key Uncertainties and Further Work Required

The prediction of future climate change is critically dependent on scenarios of future anthropogenic emissions of greenhouse gases and other climate forcing agents such as aerosols. These depend not only on factors which can be addressed by the natural sciences but also on factors such as population and economic growth and energy policy where there is much uncertainty and which are the concern of the social sciences. Natural and social scientists need to cooperate closely in the development of scenarios of future emissions.

Since the 1990 report there has been a greater appreciation of many of the uncertainties which affect our predictions of the timing, magnitude and regional patterns of climate change. These continue to be rooted in our inadequate understanding of:

- sources and sinks of greenhouse gases and aerosols and their atmospheric concentrations (including their indirect effects on global warming)
- clouds (particularly their feedback effect on greenhouse-gas-induced global warming, also the effect of aerosols on clouds and their radiative properties) and other elements of the atmospheric water budget, including the processes controlling upper level water vapour
- oceans, which through their thermal inertia and possible changes in circulation, influence the timing and pattern of climate change
- polar ice sheets (whose response to climate change also affects predictions of sea level rise)
- land surface processes and feedbacks, including hydrological and ecological processes which couple

regional and global climates

Reduction of these uncertainties requires:

- the development of improved models which include adequate descriptions of all components of the climate system
- improvements in the systematic observation and understanding of climate-forcing variables on a global basis, including solar irradiance and aerosols
- development of comprehensive observations of the relevant variables describing all components of the climate system, involving as required new technologies and the establishment of data sets
- better understanding of climate-related processes, particularly those associated with clouds, oceans and the carbon-cycle
- an improved understanding of social, technological and economic processes, especially in developing countries, that are necessary to develop more realistic scenarios of future emissions
- the development of national inventories of current emissions
- more detailed knowledge of climate changes which have taken place in the past
- sustained and increased support for climate research activities which cross national and disciplinary boundaries; particular action is still needed to facilitate the full involvement of developing countries
- improved international exchange of climate data.

Many of these requirements are being addressed by major international programmes, in particular by the World Climate Research Programme (WCRP), the International Geosphere Biosphere Programme (IGBP) and the Global Climate Observing System (GCOS). Adequate resources need to be provided both to the international organization of these programmes and to the national efforts supporting them if the new information necessary to reduce the uncertainties is to be forthcoming. Resources also need to be provided to support on a national or regional basis, and especially in developing countries, the analysis of data relating to a wide range of climate variables and the continued observation of important variables with adequate coverage and accuracy.

ANNEX

Progress in the Development of an IPCC Methodology for National Inventories of Net Emissions of Greenhouse Gases

Scientific assessment is primarily concerned with sources and sinks at the global, and large region level but, in order to support national and international responses to climate change, it is necessary to estimate emissions and sinks at the national level in an agreed and consistent way.

IPCC (1991) has established a work programme to:

(i) develop an approved detailed methodology for calculating national inventories of greenhouse gas emissions and sinks

(ii) assist all participating countries to implement this methodology and provide results by the end of 1993.

This programme is based on preliminary work sponsored by the Organization for Economic Cooperation and Development (OECD, 1991). OECD and the International Energy Agency (IEA) are continuing to provide technical support to the IPCC work programme. The programme will manage the development and approval of inventory methods and procedures, and the collection and evaluation of data. It will collaborate with other sponsors including the Global Environment Facility (GEF), the Asian Development Bank, the European Community, UNECE and individual donor countries, to encourage funding of technical cooperation projects in greenhouse gas (GHG) inventories.

The IPCC requested that participating countries provide any available GHG emissions inventory data to the IPCC by the end of September 1991. As of January 1992, 18 countries have submitted complete or partial GHG inventories (see Table below); most relate to average emissions over one, two or three years in the period 1988 to 1990. This process has been particularly useful in identifying problems in coverage and consistency of currently available inventories.

An IPCC Workshop on National GHG Inventories, held in Geneva from 5 to 6 December 1991, provided guidance on needed improvements in the draft methodology and priorities for the work programme. Numerous improvements to the methodology were agreed upon, and priorities were proposed for the work programme and for technical cooperation activities. As a result of the preliminary data collection, the workshop, and other comments received, the following major priorities for the IPCC work programme have been established:

Methodology

- Develop a simpler methodology and streamlined "workbook" document to assist users in its implementation
- Work with experts to develop a new and simpler method for calculating CO_2 emissions from forestry and land-use change
- Establish technical expert groups to improve the methodology for CH_4 from rice and fossil fuel production, and other key gases and source types
- Work with experts to include halocarbons in the GHG inventory starting with data available from the Montreal Protocol process
- Develop and disseminate regionally-applicable emissions factors and assumptions.

Annex Table: Countries which have submitted complete or partial inventories of national greenhouse gas emissions (by January 1992)

Australia	Germany	Sweden
Belgium	Italy	Switzerland
Canada	Netherlands	Thailand
Denmark	New Zealand	Vietnam
Finland	Norway	United Kingdom
France	Poland	United States

Work Programme

- Priorities for national inventories are: (a) CO_2 from energy for all countries, (b) CO_2 from forestry and land use if important for the country, and (c) CH_4 for important source categories by country
- Initiate intercomparison studies of existing detailed inventories
- Include a scientific review of national inventory data and aggregated totals by region and globally in the work programme.

Technical Cooperation

- IPCC will improve communications among technical focal points in all participating countries and with other interested international organizations
- High priority should be placed on country case studies, training, regional cooperation and other activities to assist non-OECD countries in testing and implementing the GHG inventory methodology
- Provide methods in the form of a streamlined workbook in several languages. A user-friendly computer spreadsheet version will also be developed as a high priority.

References

IPCC, 1991: Report of the Fifth Session of the WMO/UNEP Intergovernmental Panel on Climate Change (IPCC) 13-15 March 1991, Geneva.

OECD, 1991: Estimation of Greenhouse Gas Emissions and Sinks: Final Report from the OECD Experts Meeting, 18-21 February 1991, Paris. Revised August 1991.

A

Greenhouse Gases

Coordinating Author:
R.T. WATSON

A1

Greenhouse Gases: Sources and Sinks

R.T. WATSON, L.G. MEIRA FILHO, E. SANHUEZA, A. JANETOS

Contributors:
P.J. Fraser[†]; I.J. Fung; M.A.K. Khalil; M. Manning; A. McCulloch; A.P. Mitra;
B. Moore; H. Rodhe; D. Schimel; U. Siegenthaler; D. Skole; R.S. Stolarski[†]

† - Chairs for Chapters 1 and 2 of WMO/UNEP Science Assessment of Ozone Depletion: 1991 (WMO, 1992), therefore
representing a number of other contributors who are not listed here.

CONTENTS

EXECUTIVE SUMMARY

There have been no major changes in our understanding of greenhouse gases since the 1990 Scientific Assessment for the Intergovernmental Panel on Climate Change (IPCC, 1990). While most of the key uncertainties identified in IPCC (1990) remain unresolved, there have been a number of important advances.

Atmospheric Concentrations and Trends of Long-lived Greenhouse Gases:

The atmospheric concentrations of the major long-lived greenhouse gases (carbon dioxide (CO_2), methane (CH_4), nitrous oxide (N_2O), chlorofluorocarbons (CFCs), and carbon tetrachloride (CCl_4) continue to increase because of human activities. While the growth rates of most of these gases have been steady or increasing over the past decade, that of CH_4 and some of the halocarbons has been decreasing. The rate for CH_4 has declined from about 20 ppbv/yr in the late 1970s to possibly as low as 10 ppbv/yr in 1989.

Atmospheric Concentrations and Trends of Other Gases that Influence the Radiative Budget:

Ozone (O_3) is an effective greenhouse gas in the upper troposphere and lower stratosphere. Significant decreases in total column O_3 have been observed throughout the year during the last one to two decades, and at all latitudes except the tropics, with the trends being larger during the 1980s than during the 1970s. These decreases have occurred predominantly in the lower stratosphere, below 25km, where the rate of decrease has been up to 10% per decade depending on altitude. In addition, there is evidence from the few existing ozonesonde stations that, at northern mid-latitudes, O_3 levels in the troposphere up to 10km altitude have increased by about 10% per decade over the past two decades. Also, the abundance of carbon monoxide (CO) appears to be increasing in the Northern Hemisphere at about 1% per year. There is little new information on the global trends of other tropospheric O_3 precursors (non-methane hydrocarbons (NMHC) and oxides of nitrogen (NO_x)).

Sources and Sinks of Carbon Dioxide:

The best estimate for global fossil fuel emissions in 1989 and 1990 is 6.0±0.5 GtC, compared to 5.7±0.5 GtC in 1987. The estimated total release of CO_2 from oil well fires in Kuwait during 1991 was 0.065 GtC. The United Nations Food and Agriculture Organization (FAO), using information supplied by individual countries, recently estimated that the rate of global tropical deforestation in closed and open canopy forests for the period 1981-1990 was about 17 million hectares (Mha) per year, approximately 50% higher than that for the period 1976-1980. A comprehensive, multi-year, high spatial resolution satellite data set has been used to estimate that the average rate of deforestation in the Brazilian Amazonian forest between 1978 and 1989 was 2.1 Mha/yr, and that the rate increased between 1978 and the mid-1980s, and decreased to 1.4 Mha/yr in 1990. Despite the new information suggesting higher rates of deforestation globally, the uncertainties in estimating CO_2 emissions are so large that there is no convincing reason to revise the IPCC (1990) estimate of annual average net flux to the atmosphere of 1.6±1.0 GtC from land-use change during the decade of the 1980s.

Based on models and the atmospheric distribution of CO_2, it appears that: (i) there is a small net addition of carbon to the atmosphere from the equatorial region, a combination of outgassing of CO_2 from warm tropical waters and a terrestrial biospheric component that is the residual between large sources (deforestation) and sinks; (ii) there is a strong Northern Hemisphere sink, containing both oceanic and terrestrial biospheric components, and a weak Southern Hemisphere sink; (iii) an ocean sink of 2.0±0.8 GtC per year is still reasonable; and (iv) terrestrial biospheric processes are sequestering CO_2 due to forest regeneration, and fertilization arising from the effects of both CO_2 and nitrogen (N).

Sources and Sinks of Methane:

A total annual emission of CH_4 of about 500Tg can be deduced from the magnitude of its sinks combined with its rate of atmospheric accumulation. New information includes a revised rate of removal of CH_4 by atmospheric hydroxyl (OH) radicals (because of a lower rate constant), a new evaluation of some of the sources (e.g., from rice fields) and the addition of new sources (e.g., animal and domestic waste). Recent CH_4 isotopic studies suggest that approximately 100 $TgCH_4$ is of fossil origin, largely from the coal, oil, and natural gas industries. Recent studies of CH_4 emissions from rice agriculture show that the emissions depend on growing conditions, vary significantly, and are significantly lower than reported in IPCC (1990). The latest estimate of the atmospheric lifetime of CH_4 is about 11 years.

Sources and Sinks of Nitrous Oxide:

Adipic acid (nylon) production, nitric acid production and automobiles with three-way catalysts have been identified as possibly significant anthropogenic global sources of nitrous oxide. However, the sum of all known anthropogenic and natural sources is still barely sufficient to balance the calculated atmospheric sink or to explain the observed increase in the atmospheric abundance of N_2O.

Sources of Halogenated Species:

The worldwide consumption of CFCs 11, 12, and 113 is now 40% below 1986 levels, substantially below the amounts permitted under the Montreal Protocol. Further reductions are mandated by the 1990 London Amendments to the Montreal Protocol. As CFCs are phased out, HCFCs and HFCs will substitute, but at lower emission rates.

Stratospheric Ozone:

Even if the control measures of the 1990 London amendments to the Montreal Protocol were to be implemented by all nations, the abundance of halocarbons will increase over the next several years. As the weight of evidence suggests that these gases are responsible for the observed reductions in stratospheric O_3, the rate of depletion at middle and high latitudes is predicted to increase during the 1990s.

Precursors of Tropospheric Ozone:

Little new information is available regarding the tropospheric ozone precursors (CO, NMHC, and NO_x), all of which have significant, but uncertain, natural and anthropogenic sources.

Aerosols:

Industrial activity, biomass burning, volcanic eruptions, and sub-sonic aircraft contribute substantially to the formation of tropospheric and stratospheric aerosols. Industrial emissions are especially important to the tropospheric burden of aerosols; the Northern Hemisphere is particularly affected but so are any regions having a concentration of industrial activity. Sulphur emissions, which are due in large part to combustion effluents, have a similar emissions history to that of anthropogenic CO_2. Estimates of emissions of natural sulphur compounds have been reduced from previous figures, thereby placing more emphasis on the anthropogenic contribution.

A1.1 Introduction

This section updates Chapter 1 of the 1990 IPCC Scientific Assessment. The key issues are to quantify the atmospheric distributions, trends, sources and sinks of greenhouse gases and their precursors, and to understand the processes controlling their global budgets in order to relate their emissions quantitatively to atmospheric concentrations.

A1.2 Carbon Dioxide

A1.2.1 Introduction

The atmospheric accumulation of CO_2 is the balance between fossil fuel and land-use change emissions, and the uptake due to oceanic and terrestrial sinks. The key issue is to understand the processes controlling the global carbon budget in order to relate anthropogenic emissions of CO_2 quantitatively to atmospheric concentrations. Two concerns have dominated scientific debate since the first IPCC Science Assessment: (a) the global rate of land-use change, especially deforestation, and (b) the fluxes of carbon and the processes controlling its release and uptake in both the terrestrial biosphere and the oceans.

A1.2.2 Atmospheric Abundances and Trends

IPCC (1990) assigned a value of 353 ppmv to the global annual average concentration of CO_2 in 1990, based on an extrapolation of the most recent measurements available at that time. For 1991, the best global estimate is approximately 355 ppmv, given the recent observed rate of increase of 1.8 ppmv/yr. There is a small, but coherent gradient in concentration from the South Pole to the Arctic basin of about 3 ppmv (Keeling *et al.*, 1989a; Heimann *et al.*, 1989), that depends on the distribution of both emissions from fossil fuel combustion and the distribution of terrestrial and oceanic sources and sinks (IPCC, 1990). The gradient implies a continuous flux of CO_2 from the Northern Hemisphere, where about 90% of the fossil fuel emissions occur, to the Southern Hemisphere, where part of the net uptake by the oceans takes place. The concentrations of CO_2 and its stable isotope $^{13}CO_2$ show clearly that the wide and regular seasonal variability at most stations, with substantially higher amplitudes in the Northern Hemisphere, is dominated by the activity of the terrestrial biosphere in the Northern Hemisphere, rather than by seasonal changes in ocean pCO_2 or fossil sources.

The atmospheric adjustment time (IPCC, 1990) of CO_2 depends on the different time constants of many processes. A rather rapid adjustment takes place between the atmosphere, the surface layer of the oceans, and the terrestrial biosphere when anthropogenic CO_2 is added to or removed from the atmosphere. However, the long-term response of the atmospheric concentration of CO_2 to anthropogenic emissions depends primarily on the processes that control the rate of storage of CO_2 in the deep ocean and in forest biomass and soil organic matter, which have characteristic time-scales of several decades to centuries.

A1.2.3 Sources

Updates of emissions from fossil fuel combustion, including oil wells set alight in Kuwait, and changes in land-use are summarized below.

A1.2.3.1 Combustion

Marland and Boden (1991) have recently updated their estimates of CO_2 industrial emissions to include data through 1989. The updated best estimate for the global emission in both 1989 and 1990 is 6.0±0.5 GtC, compared to 5.7 GtC in 1987 (IPCC, 1990); (GtC: 1 Gt = 10^9t = 10^{15}g; 1 ppmv CO_2 of the global atmosphere = 2.12 GtC and 7.8 Gt CO_2). The estimates rely primarily on energy data published by the United Nations with supplemental data on cement production from the US Bureau of Mines and on gas flaring from the US Department of Energy.

In recent months, concerns have been raised over the possible magnitude of emissions from oil well fires lit at the end of the war in Kuwait. About 600 naturally pressurized oil wells were set alight in Kuwait in late February 1991, but by the beginning of November all the wells had been capped. The pre-war production rate of Kuwait was about 1.6 million barrels per day (MBd^{-1}). Airborne measurements of the chemical composition of the plumes were made in late March 1991 (Johnson *et al.*, 1991); the major pollutants observed were particulates (smoke), CO_2, sulphur dioxide (SO_2), nitrogen oxides (NO_x), and unburnt hydrocarbons. Based on the sulphur composition of the oil, the oil burning rate was estimated to be equivalent to 3.9±1.6 million barrels per day. The estimated total emissions of carbon (C) as CO_2 during 1991 was 65Tg, or about 1% of the estimated annual global fossil fuel emissions; uncertainty in this figure is approximately ±50%.

A1.2.3.2 Land-Use Changes

The net flux of C to the atmosphere from land-use change (primarily, though not exclusively due to deforestation in the tropics) depends on the area converted, carbon density per ha, the fate of the altered land and the ecosystem processes that control fluxes of carbon. The IPCC (1990) estimate for the flux in 1980 was 0.6-2.5 GtC, with suggestions that the annual flux was higher in 1990 than in 1980. Houghton (1991) has recently calculated that the release of C to the atmosphere from land-use change in 1990 was 1.1-3.6 GtC, somewhat higher than the 1980 estimate of 0.6-2.5 GtC. The IPCC (1990) estimate of net average annual emissions for the decade 1980-1989 was 1.6±1.0 GtC, which is consistent with Houghton's figure within the limits of uncertainty.

Table A1.1: *Summary of Global Annual Deforestation Estimates (10^6 ha)*

Reference	Myers 1980	FAO/UNEP 1981	Myers 1989	FAO 1991	WRI 1990
Year of Deforestation	1979	1976-1980	1989	1981-1990	late 1980's
Closed Canopy Forest Only	7.3	7.3	13.9	14.0	16.5
Closed and Open Canopy Forests	-	11.3	-	17.0	20.4

Table A1.2: *Summary of Brazilian Annual Deforestation Estimates (10^6 ha)*

Reference	FAO/UNEP 1981	Myers 1989	WRI 1990	Fearnside et al. 1990		INPE 1991
Year of Deforestation	1976-1980	1989	late 1980s	1978-1989	1987-1988	1989
Estimate	1.4	5.0	8.0	2.1	1.9	1.4

A1.2.3.2.1 Recent estimates of deforestation rates

The increase in Houghton's net flux estimate was primarily due to the use of Myers' (1989) new estimate of global deforestation in closed canopy forests in the late 1980s (13.9 Mha/yr for 1989), which is 90% higher than the earlier estimate (Myers, 1980) of 7.3 Mha/yr for the late 1970s. In addition, the World Resources Institute (WRI, 1990) estimated global deforestation in closed canopy forests in the late 1980s at 16.5 Mha/yr, and there is a preliminary estimate of 14.0 Mha/yr during the 1980s by FAO (1990, 1991) based on country estimates. The Myers (1989) value assumed that the rate of deforestation in Brazil was 5.0 Mha/yr, while the WRI (1990) value assumed a rate of deforestation in Brazil of 8.0 Mha/yr. The assumed rates of deforestation in Brazil were based on work of Setzer and Pereira (1991), who used saturated thermal infrared AVHRR Channel 3 data to estimate the rate of deforestation in 1987 (8.0 Mha/yr) and 1988 (5.0 Mha/yr). However, the results of Setzer and Pereira (1991) are now thought to be incorrect (Fearnside *et al.*, 1990) because of inadequate spatial resolution, and questionable assumptions about the relationship between the rate of forest clearing and the incidence of fires. Fearnside *et al.*, (1990) used comprehensive LANDSAT high-spatial resolution data sets from 1978, 1987, 1988, and 1989 to estimate that the mean rate of deforestation in the Brazilian Amazonian forest was 2.1 Mha/yr between 1978 and 1989, and that the rate from 1987 (1/3 data)/1988 (2/3 data) to 1989 was 1.9 Mha/yr. INPE (1991) has recently included data for 1990 and concluded that the rate of deforestation from 1989 to 1990 was 1.4 Mha/yr. It appears that the rate of deforestation in Brazil reached a peak sometime in the second half of the 1980s. An ongoing analysis of "wall-to-wall" LANDSAT data for 1975, 1978, 1985, and 1988-1991 should provide a much improved historical record of deforestation in Brazil. In addition, similar data for the period 1985 to 1991 is currently being analysed for all of tropical South America. The Myers (1989) and WRI (1990) global estimates of deforestation could both be revised downward to approximately 10.5 to 11.0 Mha/yr assuming that the actual rate of deforestation in Brazil was close to 2 Mha/yr for the late 1980s. Tables A1.1 and A1.2 summarize estimates of deforestation globally and in Brazil. Each of these studies used different methodologies, covered slightly different time periods, and the difference in the estimated increase in rates between the late 1970s and late 1980s is unresolved.

FAO (1990, 1991) has released a preliminary estimate, based on country submissions, of global deforestation in closed and open canopy forests for the period 1981-1990 of 17 Mha/yr, 50% higher than their estimate of 11.3 Mha/yr for 1976-1980. However, FAO acknowledges that some of the apparent increase may be due to underestimates of deforestation in the earlier time period. FAO has indicated that a rate of 6.8 Mha/yr has been estimated for tropical South America, and that the rate assumed for the Brazilian Amazonian forest was that reported by Fearnside *et al.*, (1990), i.e., 2.1 Mha/yr. This seems to ascribe a very high proportion (about 70%) of the total deforestation in South America in the late 1980s to the region outside of the Brazilian Amazonia, even though this region accounts for very little of the total amount of forest on the continent.

It seems likely that the area deforested globally in 1990 was higher than the area deforested in 1980, since the major global estimates show increases in rates of 50% (FAO, 1990, 1991) to 90% (Myers, 1989). However, the data quality for southern Asia and Africa are poor, and the results for tropical South America, which strongly influence the global estimates, are at the same time the most intensively studied and yet still uncertain. There are no global estimates of deforestation in the peer-reviewed scientific literature derived from studies using common methodologies, especially using high spatial resolution satellite data, such as LANDSAT and SPOT, which have the most potential for resolving this issue.

A1.2.3.2.2 Carbon density and ecosystem processes

Critical variables to take into account for calculating the net carbon released from deforestation and its distribution among CO_2, CO and CH_4 include the relative contribution of above-ground and below-ground carbon; the distribution between immediate releases and the effects of deforestation on the processes controlling the subsequent fluxes of carbon; the fate of deforested land; and the final carbon density of the land compared to its original carbon density. Several of these factors are poorly known locally, and depend on the sampling methods used. For example, carbon density values based on ecological data from small plots generally provide higher estimates than those from forest inventories (e.g., 328 Mg biomass/ha versus 175 Mg/ha for tropical closed forests, Brown and Lugo, 1984). All the factors are poorly known globally. Such problems must be reconciled in order to make reasonable estimates on global scales (Brown and Lugo, 1991).

In situ degradation of forest stands by selective removal of the largest or most valuable trees also affects net flux from deforestation, since the biomass at deforestation is lower than for undisturbed stands (Lugo and Brown, 1992). Other environmental factors have the potential to influence carbon density or ecosystem processes controlling carbon fluxes. For example, recovery from fires is accompanied by increasing biomass, but damage from air pollution may decrease biomass and reduce carbon fixation. However, there is no strong evidence to suggest that the frequency, intensity, or global significance of these factors has changed over the past several decades.

A1.2.3.2.3 Annual average fluxes

The estimates of average annual net flux of carbon released to the atmosphere from land-use change still appear to be among the most uncertain numbers in the carbon cycle. Houghton's (1991) estimate for 1990 suggests that the average annual net flux for the 1980s is higher than the IPCC's previous estimate. However, measurement uncertainties in the total area deforested and its average carbon density prevent accurate estimates of

changes in average annual net fluxes from being made.

The conclusion is that there is no strong evidence to revise the IPCC (1990) estimate of annual average net flux to the atmosphere from land-use change during the decade of the 1980s. The likeliest scenario is that annual fluxes increased during the early part of the decade, peaked in the middle years, and may have fallen since, based on our best understanding of deforestation rates in closed forests. While this state of affairs is unsatisfying, it is within the range of measurement error, and is representative of the present degree of scientific uncertainty.

A1.2.4 Sinks

This section summarizes new information on sinks of carbon in the oceans and the terrestrial biosphere.

A1.2.4.1 Oceans

The IPCC (1990) estimate of the global annual ocean sink, 2.0±0.8 GtC, is still reasonable. Global ocean sinks in the neighbourhood of 1.0 GtC/yr or smaller (Tans et al., 1990) appear less likely, but are still within measurement error. Estimates of the strength of the ocean sink depend on air-sea exchange, which is a function of wind speed and temperature, the difference in the partial pressure of CO_2 between the water and the air ($\sim pCO_2$), and on the initial assumptions of the ocean models. A quantitatively significant role for carbon transport by rivers would imply that ocean uptake estimates based primarily on $\sim pCO_2$ will be underestimates.

Direct measurements of the relative decline in $^{13}C/^{12}C$ ratios in dissolved inorganic carbon in ocean waters and in atmospheric CO_2 yield an estimate of 2.3 GtC/yr for the global ocean sink for the period between 1970 and 1990 (Quay et al., 1992).

While widespread episodes of coral bleaching are of regional concern for the integrity of marine ecosystems, there is no evidence that they are influencing the ocean's carbon budget in any significant way. Their abundance and distribution in the world's oceans compared to the total amount of carbon stored in the ocean is far too small to be significant.

A1.2.4.2 Terrestrial Biosphere

A major issue reported in IPCC (1990) was due to the Tans et al. (1990) analysis, which suggested a terrestrial sink in the north temperate latitudes, of the order of 1.6-2.4 GtC/yr during the decade of the 1980s. This result was obtained through inversion of the observed atmospheric CO_2 distribution combined with an atmospheric tracer model from which was derived a distribution of sources and sinks of carbon at the surface, constrained by oceanic pCO_2 data. The IPCC analysis used a value of 1.6±1.4 GtC/yr as the net imbalance in the carbon cycle during the decade of the 1980s, but did not specify whether it might be found in the terrestrial biosphere or in the oceans. The Tans and IPCC analyses have received considerable attention over the last year, primarily because of the difficulty of accounting simultaneously for very large carbon sinks in the terrestrial biosphere in northern temperate latitudes and a very small global oceanic sink.

Keeling *et al.* (1989a,b), Heimann *et al.* (1989), Heimann and Keeling (1989), Heimann (1991), Enting and Mansbridge (1991), and Tans *et al.* (1990) agree that a strong Northern Hemisphere sink and an apparently weak Southern Hemisphere sink must exist. The Northern Hemisphere sink must contain both oceanic and terrestrial biospheric components. The models disagree on the meridional distribution of total sources and sinks, and the allocation of those sinks to oceans and the terrestrial biosphere (Heimann, 1991). Tans *et al.* (1990) partitioned the northern sink of carbon between land and ocean, based on the data compiled on pCO_2 in surface waters. Recent measurements have shown large CO_2 drawdown in spring blooms in the North Atlantic (Watson *et al.*, 1991; Kempe and Pegler, 1991). On the other hand, Murphy *et al.* (1991) found less CO_2 flux into the South Pacific Ocean in austral spring than estimated by Tans *et al.* (1990). How these regional estimates combine to give an improved globally and annually averaged net flux between the atmosphere and ocean is still to be resolved. Keeling *et al.* (1989b) concluded that a large fraction of the Northern Hemispheric sink is due to a natural imbalance in the oceanic carbon cycle, consisting of a net transfer from the Northern to Southern Hemispheres, balanced by a return flux through the atmosphere, suggesting a much smaller northern terrestrial sink than calculated by Tans *et al.* (1990).

Models and atmospheric CO_2 concentration data suggest that there is a relatively small net addition of carbon to the atmosphere from the equatorial region, which is a combination of outgassing of carbon from warm tropical waters and a biospheric component. While land-use change in the tropics is an increasing and significant source of CO_2 to the atmosphere, Keeling *et al.* (1989a,b) conclude that about 50% of the terrestrial sink due to CO_2 fertilization may occur in this region, thus reducing the net flux of CO_2 from this region. These conclusions need to be tested, but it is clear that the net carbon added to the atmosphere in the tropics is less than expected simply from deforestation.

The likeliest terrestrial biospheric processes contributing to large sinks are enhancement in productivity due to atmospheric CO_2 increases, N fertilization from atmospheric deposition, and forest regrowth. Enhancement of photosynthesis, reduction in respiration, and increased water-use efficiency due to enhanced CO_2 concentrations have all been found in laboratory and field experiments (Bazzaz, 1990). Quantifying a net ecosystem response in terms of carbon storage is more difficult, and no field studies of natural forests have yet been done. Experiments in wetlands (Drake and Leadley, 1991) show net C accumulation. Analogous experiments in Arctic tundra show contrasting results, from acclimation due to the variability of the effects of CO_2 on individual species (Prudhomme *et al.*, 1984; Oechel and Strain, 1985; Tissue and Oechel, 1987) to a small net accumulation (Grulke *et al.*, 1990). The conclusion is that while the individual effects of CO_2 enhancement on plant growth and physiology are well-documented, the net ecosystem consequences of CO_2 increases under natural conditions depend on many other factors and cannot currently be quantified. Nevertheless, in models with active biospheres, global estimates of net carbon accumulation, from the physiological effects of increased CO_2 are often made in order to account for a missing sink.

Fertilization of the northern temperate latitudes through increased N deposition due to air pollution and increased fertilizer use is possibly of sufficient magnitude to sequester an additional 1.0 GtC/yr during the decade of the 1980s (IPCC, 1990; Thornley *et al.*, 1991). However, the N deposition data are poorly known, and the extent of any increase in growth rates of forests and grasslands must be weighed against the possibility of other pollutants adversely affecting physiological processes, either directly (e.g., O_3 pollution) or indirectly (acid deposition affecting soil nutrients over decades). The possible adverse effects, while suspected to be important on regional scales in eastern North America and Europe, are insufficiently quantified to provide global estimates.

Forest regrowth in temperate latitudes may currently be responsible for net C accumulation in the terrestrial biosphere. Regeneration seen in such areas as the eastern US, Europe and China often reflects recovery of land previously deforested or degraded in the late nineteenth to mid-twentieth centuries.

A1.2.5 Models and Predictions

While there are a variety of carbon cycle models, including 3-D ocean-atmosphere models, 1-D ocean-atmosphere box-diffusion models, and box models that incorporate a terrestrial biospheric sink, all such models are subject to considerable uncertainty because of an inadequate understanding of the processes controlling the uptake and release of CO_2 from the oceans and terrestrial ecosystems. Some models assume a net neutral terrestrial biosphere, balancing fossil fuel emissions of CO_2 by oceanic uptake and atmospheric accumulation; others achieve balance by invoking additional assumptions regarding the effect of CO_2 fertilization on the different parts of the biosphere. However, even models that balance the past and contemporary carbon cycle will not predict future atmospheric concentrations accurately unless they represent the proper mix of processes on land and in the oceans, and how these processes will evolve along with changes in atmospheric CO_2 and climate. For a given emissions scenario, the differences in predicted changes in CO_2 concentrations, neglecting biospheric feedbacks, are up to 30%, but this is unlikely to represent the major uncertainty in the prediction of future climate change compared to uncertainties in estimating future patterns of trace gas emissions, and in quantifying physical climate feedback processes. Future atmospheric CO_2 concentrations resulting from given emissions scenarios may be estimated by assuming that the same fraction remained airborne as has been observed during the last decade, i.e., 46±7%.

Great strides have been made in developing models that adequately simulate the terrestrial carbon budget. Regional models that simulate the response of ecosystems to CO_2 and N fertilization now exist and are being implemented for global studies (Raich *et al.*, 1991; Parton *et al.*, 1987; Thornley *et al.*, 1991). The current challenge lies more in the development of data sets adequately describing the heterogeneity of natural ecosystem types and land use, and of techniques for validation over large areas. Improvement in the estimates of present and future global ocean uptake will require better understanding of the contribution of high-frequency pCO_2 fluctuations to the global mean and a better description of the vertical mixing rate in the ocean.

Table A1.3: *Estimated Sources and Sinks of Methane (Tg CH_4 per year)*

Sources		
Natural		
• Wetlands	115	(100-200)
• Termites †	20	(10-50)
• Ocean	10	(5-20)
• Freshwater	5	(1-25)
• CH_4 Hydrate	5	(0-5)
Anthropogenic		
• Coal Mining, Natural Gas & Pet. Industry †	100	(70-120)
• Rice Paddies †	60	(20-150)
• Enteric Fermentation	80	(65-100)
• Animal Wastes †	25	(20-30)
• Domestic Sewage Treatment †	25	?
• Landfills †	30	(20-70)
• Biomass burning	40	(20-80)
Sinks		
Atmospheric (tropospheric + stratospheric) removal †	470	(420-520)
Removal by soils	30	(15-45)
Atmospheric Increase	32	(28-37)

† indicates revised estimates since IPCC 1990

A1.3 Methane

Methane (CH_4) is an important greenhouse gas. Chemical reactions involving CH_4 in the troposphere can lead to ozone production, and reaction with the hydroxyl radical (OH) in the stratosphere results in the production of water vapour. Both tropospheric ozone, especially in the upper troposphere, and stratospheric water vapour are significant greenhouse gases. The final oxidation product of CH_4 is CO_2 which is also a greenhouse gas. One of the major removal mechanisms for OH is reaction with CH_4, therefore, as the levels of CH_4 increase the levels of OH could decrease. This could lead to increased lifetimes of CH_4 and other important greenhouse gases, e.g., hydrochlorofluorocarbons (HCFCs).

A1.3.1 Atmospheric Concentrations and Trends

The present atmospheric concentration of CH_4 is 1.72 ppmv globally averaged, more than double its pre-industrial value of about 0.8 ppmv. The abundance in the Northern Hemisphere is almost 0.1 ppmv greater than in the Southern Hemisphere. Recent data, which were carefully reviewed in WMO (1992), verify that the rate of growth of the atmospheric concentration of CH_4 slowed during the last decade or so from a rate of about 20 ppbv per year (about 1.3% per year: Blake and Rowland, 1988)

in the late 1970s, to a rate of about 12-13 ppbv per year (about 0.75% per year) in the mid-1980s, to possibly as low as 10 ppbv per year (about 0.6% per year) in 1989 (Steele *et al.*, 1992). A range of 10-13 ppbv per year is reported in Table A1.3. There are no convincing explanations for this decline in CH_4 growth rates. It could be due to a decrease in emission rates from natural or anthropogenic sources, an increase in its loss rate due to an increase in the concentration of tropospheric OH, which has been suggested by Prinn *et al.*, (1992), or a combination of the two.

A1.3.2 Sinks

The main sink of CH_4 is reaction with the OH radical in the troposphere, so any estimate of the magnitude of this sink requires knowledge of the rate constant and the atmospheric abundance of OH. Current estimates of lifetime and global sources must be viewed with caution until more reliable values of the atmospheric abundances of OH are derived. Vaghjiani and Ravishankara (1991) have shown that the rate constant for CH_4 reacting with OH had been overestimated by up to 25%. Using this new rate constant data, Crutzen (1991) and Fung *et al.*, (1991) estimated the annual current removal of CH_4 by OH to be

420 ± 80 TgCH$_4$. Crutzen (1991) also estimated an additional sink of 10 ± 5 TgCH$_4$ due to photochemical removal in the stratosphere. Prinn *et al.*, (1992) have estimated higher atmospheric concentrations of OH using the ALE/GAGE methyl chloroform (CH$_3$CCl$_3$) data, and with the new OH + CH$_4$ rate constant data have deduced a lifetime for CH$_4$ of 11.1 (+1.4, -1.1) years and a net CH$_4$ source of 470 ± 50Tg per year for the period 1978-1990. The Prinn *et al.* (1992) estimate of lifetime, and hence of source strength, includes both tropospheric and stratospheric chemical removal. The annual magnitude of an additional sink, i.e., uptake by soils, has been estimated to be 30 ± 15 TgCH$_4$ (IPCC, 1990; Whalen and Reeburg, 1990). Recent CH$_4$ soil flux measurements indicate that changes in land use (Keller *et al.*, 1990; Scharffe *et al.*, 1990) or enhanced nitrogen fertilizer input (Mosier *et al.*, 1991) are decreasing the CH$_4$ uptake by soils. While the current magnitude of the soil sink is relatively small, the importance of this sink may change in the future if changes in climatic conditions result in significant changes in soil moisture. The lifetime of CH$_4$ is presently estimated to be about 11 ± 2 years.

A1.3.3 Sources

A total annual global emission of CH$_4$ of about 500 TgCH$_4$ can be deduced from the magnitude of its sinks and its rate of accumulation in the atmosphere. Although the emission rates from most of the individual sources are still quite uncertain, there is a fairly good balance between the sum of the individual sources and the deduced global emission rate. The natural and anthropogenic sources of CH$_4$ were discussed in detail in IPCC (1990), and in most cases the current estimated global source strengths are unchanged. Table A1.3, which summarizes the best current estimates of emissions from individual sources, clearly shows that anthropogenic sources dominate ($\sim$2:1) over natural sources, consistent with a more than doubling of the atmospheric abundance of CH$_4$ since pre-industrial times. There has been a revaluation of some of the sources, particularly from rice, and the addition of new sources such as animal and domestic waste.

The proportion of CH$_4$ produced from sources related to fossil carbon can be estimated from studies of the carbon-14 content of atmospheric CH$_4$. Three independent estimates are: $21\pm3\%$ (Whalen *et al.*, 1989), $25\pm4\%$ (Manning *et al.*, 1990) and $16\pm12\%$ (Quay *et al.*, 1991).

Table A1.4: *Measured Methane Emissions During the Growing Season From Rice Paddies*

	Flux mg m^{-2} h^{-1}	Annual rate g m^{-2} yr^{-1}	Comments
California [1]	10	25-42	1982 growing season
Spain [2] (Andalucia)	4	12	Affected by sulphate from sea water
Italy [3]	12 (6-16)	14-77	7 fertilization treatments 3 veg. periods
Japan [4] (Ibaraki-prefec.)	16.2	45	Peat soil
	2.9-15.4	8-43	Gley soil
	<0.4-4.2	<1-13	Andosol
China [5] (Tu Zu, Sczhuan Province)	60 (10-120)	170	4 rice fields, 6 plots each site, local fertilization practices; 2 veg. periods, range due, in part, to seasonal variation
China [6] (Hangzhou, Zhajiang Provinces)	28.6	55-97	Late rice
	7.8	14-18	Early rice
China [7] (Beijing)	15-50	-	Rice after wheat; 5 different methods of management; 1 veg. period
China [8] (Nanjing)	11(3-14)	-	5 different managements
India [9]	0.1-27.5	7.5-22.5	Includes irrigated and rainfed fields; acidic and non-acid soils
Texas [10]	2.5-8.7	5-16	-
Australia [11]	3.8	-	Micrometeorological measurements
Thailand [12]	3.7-19.6	8-42	-

(1) Cicerone and Shetter 1981; Cicerone *et al.*, 1983; (2) Seiler *et al.*, 1984; (3) Holzapfel-Pschorn and Seiler, 1986; Schutz *et al.*, 1989; (4) Yagi and Minami, 1990; (5) Khalil and Rasmussen, 1991; (6) Wang *et al.*, 1989; (7) Zongliang *et al.*, 1992; (8) Debo *et al.*, 1992; (9) Parashar *et al.*, 1991; Mitra, 1992; (10) Sass *et al.*, 1990; (11) Denmead and Freney, 1990; (12) Yagi, unpublished.

Thus, about 20% (100Tg) of total annual CH_4 emissions are from fossil carbon sources, primarily coal-mining operations, and oil and natural gas production, transmission, distribution and use. A very small part of this may also be due to release of old CH_4 from hydrate destabilization (0-5Tg), and a warmer climate could lead to a significant increase in the magnitude of this source. The latest estimates of global annual emissions for coal-mining range from 25 to 47Tg (Okken and Kran, 1989; Barns and Edmonds, 1990; Boyer *et al.*, 1990; Hargraves, 1990); for the gas industry from 25 to 42Tg (Okken and Kran, 1989; Barns and Edmonds, 1990); and for the oil industry from 5 to 30Tg (Okken and Kran, 1989). Emissions from these three sources are broadly consistent with a total annual emission of 100Tg.

Emissions from wetlands are the largest natural source of CH_4 to the atmosphere. Although the recent estimate is in the range of the IPCC (1990) estimate, the latitudinal distribution of the emissions has been revised. Emissions from high-latitude peat bogs are about half of that previously estimated, while inclusion of bubbling tends to enhance the estimates of emissions from tropical swamps.

The 1990 IPCC estimate of the CH_4 flux from rice paddies was based on a very limited amount of data. Table A1.4 summarizes CH_4 emission rates from rice cultivation, including a substantial amount of new information from Japan, Australia, Thailand and, in particular, India and China. Very large variations in emissions are observed between different rice paddies. These differences are probably due to several factors including irrigation and fertilization practices, soil/paddy characteristics (particularly redox potential), cultivation history, temperature, and season. While WMO (1992) decided to retain the same quantitative emission estimate of 100Tg as IPCC (1990), a detailed analysis of the available data, particularly that from India, suggests a global annual emission nearer the lower end of the range.

While the estimate of CH_4 emissions from landfills has been slightly reduced because of observed high CH_4 oxidation rates in landfill cover soils (Whalen *et al.*, 1990), two "waste" sources of CH_4 were not included in the previous IPCC assessment, i.e., emissions from animal waste (Casada and Safley, 1990) and domestic sewage treatment (Harriss, 1991). Based on recent studies (Khalil *et al.*, 1990; Rouland *et al.*, 1991) the annual emission from termites was scaled down to 20Tg. Carbon-13 studies (Stevens, 1988; Quay *et al.*, 1991; Lasaga and Gibbs, 1991) indicate that the contribution from biomass burning is in the upper part of the range given in Table A1.3.

A1.4 Nitrous Oxide

Nitrous oxide is an important long-lived greenhouse gas. In addition, it is the primary source of the oxides of nitrogen in the stratosphere, which play a critical role in controlling the abundance and distribution of stratospheric ozone.

A1.4.1 Atmospheric Concentrations and Trends

Numerous new data confirm that the present-day atmospheric concentration of nitrous oxide (N_2O) is about 310 ppbv; about 8% greater than during the pre-industrial era, and it is increasing at a rate of 0.2-0.3% per year (WMO, 1992). This observed rate of increase represents an annual atmospheric growth rate of about 3 to 4.5TgN.

A1.4.2 Sinks

The major atmospheric loss processes for N_2O are stratospheric photo-dissociation and stratospheric photo-oxidation. Consumption in soils may represent a small sink but, to date, this has not been evaluated. The atmospheric N_2O lifetime is about 130 (110-168) years (WMO, 1992).

A1.4.3 Sources

The sources of N_2O were previously discussed in detail in IPCC (1990). A revised budget is given in Table A1.5 based on new information on soil fluxes from tropical ecosystems (Sanhueza *et al.*, 1990; Matson *et al.*, 1990) and temperate forests (Bowden *et al.*, 1990); detailed evaluations of cultivated soils (Bouwman, 1990; Eichner, 1990); and new estimates of emissions from biomass burning (Lobert *et al.*, 1990; Cofer *et al.*, 1991). Large tropical sources are required to explain the N_2O latitudinal

Table A1.5: *Estimated Sources and Sinks of Nitrous Oxide (Tg N per year)*

Sources	
Natural	
• Oceans	1.4-2.6
• Tropical Soils	
• Wet forests	2.2-3.7
• Dry savannas	0.5-2.0
• Temperate Soils	
• Forests	0.05-2.0
• Grasslands	?
Anthropogenic	
• Cultivated Soils	0.03-3.0
• Biomass Burning	0.2-1.0
• Stationary Combustion	0.1-0.3
• Mobile Sources	0.2-0.6
• Adipic Acid Production	0.4-0.6
• Nitric Acid Production	0.1-0.3
Sinks	
Removal by soils	?
Photolysis in the Stratosphere	7-13
Atmospheric Increase	3-4.5

gradient determined using 10 years of ALE/GAGE N$_2$O data. The analysis shows specifically that emissions into the latitude bands 90°N-30°N, 30°N-equator, equator-30°S, and 30°S-90°S, account for about 22-24%, 32-39%, 20-29%, and 11-15%, respectively, of the global total (Prinn *et al.*, 1990). The impact of tropical deforestation on the emissions of N$_2$O is unclear; one study indicates an enhancement after conversion of humid forest to pasture, whereas other studies conclude that forest degradation (Sanhueza *et al.*, 1990) or conversion to pasture (Keller, 1992) reduces the emissions.

Several recent publications (i.e., De Soete, 1989; Linak *et al.*, 1990; Yokoyama *et al.*, 1991) reconfirm that N$_2$O emissions from stationary combustion sources are very low. Fluidized bed combustion systems produce larger quantities of N$_2$O than the traditional coal combustors (De Soete, 1989), but because of their present limited application their contribution is negligible (CONCAWE, 1991).

Several smaller, but important, anthropogenic sources of N$_2$O have now been identified. Efforts should continue to identify as yet unidentified sources. Thiemens and Trogler (1991) estimate that adipic acid production (for nylon 66) results in annual atmospheric emissions of 0.4TgN per year (industrial estimates suggest that it might be as high as O.6TgN). The annual emission from nitric acid production (mainly for N-fertilizer) is 0.1 to 0.3TgN (McCulloch, 1991). Three-way catalyst controlled vehicles have higher N$_2$O emissions than uncontrolled vehicles (De

Soete, 1989) and recent annual estimates of emissions vary between 0.2 and 0.6TgN.

While the estimated source strengths are quite uncertain, it appears that emissions from soils dominate the N$_2$O budget. Since the total annual emission rate of N$_2$O appears to be between 10 and 17.5TgN, deduced from the magnitude of its sinks and its rate of accumulation in the atmosphere, and the estimated annual sources are between 5.2 and 16.1TgN, it seems that the strengths of some of the identified sources have been underestimated or that there are unidentified sources. It has been suggested that agricultural development may stimulate biological production and account for the missing emissions. Although large changes in land use are occurring in the tropics no evaluation of the impact of the increasing use of nitrogen fertilizers has been made for this region.

A1.5 Halogenated Species
A1.5.1 Atmospheric Trends
The major halogenated source gases, i.e., CFCs, HCFC-22, the halons, methyl chloroform (CH$_3$CCl$_3$) and carbon tetrachloride (CCl$_4$), continue to grow in concentration in the background troposphere of both Hemispheres even though the consumption of most of these gases has decreased significantly in recent years (Section A1.6.2). The data are summarized in Table A1.6. The fully fluorinated species, tetrafluoromethane (CF$_4$), hexa-fluoroethane (C$_2$F$_6$), and sulphur hexafluoride (SF$_6$) have been measured in the atmosphere, with CF$_4$ being the most abundant species (70-80 pptv).

A1.5.2 Sinks
The only significant sink for the fully halogenated CFCs is photolysis in the stratosphere, whereas the primary sink for the partially halogenated chemicals is reaction with OH in the troposphere.

A1.5.3 Sources
The worldwide consumption of CFCs 11, 12 and 113 decreased by 40% between 1986 and 1991. At this rate, developed country consumption will be reduced by 50% in 1992 - a three-year advance on the requirements of the London Amendments to the Montreal Protocol on Substances that Deplete the Ozone Layer, which calls for a 50% reduction by 1995, 85% reduction by 1997, and a complete phaseout by the year 2000. Major reductions have been made by using hydrocarbons as aerosol propellants and as blowing agents for flexible foams. Solvent users are turning to aqueous and semi-aqueous systems, no-clean technologies, alcohol, and other solvents. Refrigeration and air conditioning sectors are recovering and recycling CFC refrigerants and increasing the use of HCFCs and ammonia. Insulating foams have a lower CFC content with little energy penalty.

Table A1.6: *Atmospheric Concentrations and Trends of Halocarbons*

Species	Annual Mean Concentration (pptv), 1989	Trend (pptv/yr) 1987 WMO (1989)	1989 WMO (1992)
CFC-11	255 †	9 - 10	9 - 10
CFC-12	453 †	16 - 17	17 - 18
CFC-113	64 †	5	6
CCl$_4$	107 †	2	1 - 1.5
HCFC-22	110	7	5 - 6
CH$_3$CCl$_3$	150 †, ¥	6	4 - 5
CBrClF$_2$	1.8 - 3.5	0.2	0.4 - 0.7
CBrF$_3$	1.6 - 2.5	0.3	0.2 - 0.4

† based on GAGE data (Prinn *et al.*, 1991 and references therein)

¥ the 1987 mean CH$_3$CCl$_3$ concentration was 140 pptv, not 150 pptv as reported in WMO (1989).

Table A1.7: *Total Ozone Trends in Percent per Decade with 95% Confidence Limits*

| | TOMS: 1979-1991 | | | Ground-based: 26°N - 64°N | |
	45°S	Equator	45°N	1979-1991	1970-1991
Dec-Mar	-5.2 ± 1.5	+0.3 ± 4.5	-5.6 ± 3.5	-4.7 ± 0.9	-2.7 ± 0.7
May-Aug	-6.2 ± 3.0	+0.1 ± 5.2	-2.9 ± 2.1	-3.3 ± 1.2	-1.3 ± 0.4
Sep-Nov	-4.4 ± 3.2	+0.3 ± 5.0	-1.7 ± 1.9	-1.2 ± 1.6	-1.2 ± 0.6

Emissions of halocarbons will decrease further because reductions are mandated by the 1990 London Amendments to the Montreal Protocol. In addition, many countries and some industries have called for an even faster phase-out of the controlled substances. UNEP (1991) reported that it is technically feasible to almost completely phase out controlled substances in developed countries by 1995 - 1997 through a number of measures that are technically feasible, many either economically advantageous or at no economic cost, others at a "modest" economic cost. A more rapid phase-out depends on the extent of recycling and technical feasibility of equipment retrofit, on the availability of HCFC and HFC replacements and on their toxicological and environmental acceptability, on a regulatory regime which allows profitable investment in their production, on vigorous and effective management of the halon bank, and on the very rapid dissemination and adoption of technologies for the replacement of CH_3CCl_3 by small users.

The main sources for the fully fluorinated chemicals are not well understood. Production of aluminium and use in electrical equipment are probably some of the most important sources (Stordal and Myhre, 1991).

A1.6 Ozone

Ozone (O_3) is a particularly effective greenhouse gas in the upper troposphere and lower stratosphere, and also plays a key role in absorbing solar ultraviolet radiation. About 90% of the total column of ozone resides in the stratosphere, and the remaining 10% in the troposphere. The current scientific and policy concerns are reduction of stratospheric ozone by chlorine- and bromine-containing chemicals, and production of tropospheric ozone by carbon monoxide, hydrocarbons, and oxides of nitrogen (see Section A.2.6). This section is a summary of a very extensive international scientific assessment (WMO, 1992), which should be read for a more comprehensive discussion and for all key references.

A1.6.1 Observed Trends in Total Column Ozone
Marked Antarctic O_3 holes have continued to occur and, in four of the past five years, have been deep and extensive in area. This contrasts to the situation in the mid-1980s, where the depth and area of the O_3 hole exhibited a quasi-biennial modulation. Recent laboratory research and re-interpretation of field measurements have strengthened the evidence that the Antarctic O_3 hole is primarily due to chlorine- and bromine-containing chemicals. While no extensive O_3 losses have occurred in the Arctic comparable to those observed in the Antarctic, localized Arctic O_3 losses have been observed in winter concurrent with observations of elevated levels of reactive chlorine.

Recent ground-based and satellite observations (Stolarski *et al.*, 1991) show evidence for significant decreases of total column O_3 throughout the year in both the Northern and Southern Hemispheres at middle and high latitudes (Table A1.7). No trends in O_3 have been observed in the tropics. These downward trends were larger during the 1980s than during the 1970s.

The observed decreases in total column ozone comprise a decrease in the stratosphere, possibly offset to some degree by an increase in the troposphere (see next section).

A1.6.2 Observed Trends in the Vertical Distribution of Ozone
Ground-, balloon-, and satellite-based measurements show that the observed total column O_3 decreases during the last one to two decades are predominantly due to O_3 concentration decreases in the lower stratosphere (between the tropopause and 25km altitude or lower). The rate of decrease at middle latitudes in both hemispheres, and high latitudes in the Northern Hemisphere has been up to 10% per decade depending on altitude. Ozone decreases exceeding 90% have been observed in the lower stratosphere within the springtime Antarctic ozone hole. In addition, there is evidence of small decreases globally in the upper stratosphere.

Measurements indicate that above the few existing balloonsonde stations at northern middle latitudes in Europe O_3 levels in the troposphere up to 10km altitude have increased by about 10% per decade over the past two decades. However, the data base for O_3 trends in the upper troposphere is sparse and inadequate for quantifying its contribution to the global radiative balance.

A1.6.3 *Future Levels of Stratospheric Ozone*

The weight of evidence suggests that the observed middle- and high-latitude O_3 losses are largely due to chlorine and bromine. Therefore, as the atmospheric abundances of chlorine and bromine increase in the future, significant additional losses of O_3 are expected. Even if the control measures of the amended Montreal Protocol (London, 1990) were to be implemented by all nations, the current abundance of stratospheric chlorine (3.3-3.5 ppbv) is estimated to increase during the next decade, reaching a peak of about 4.1 ppbv around the turn of the century. With these increases, additional middle-latitude O_3 losses during the 1990s are expected to be comparable to those observed during the 1980s. Hence, by the year 2000 O_3 depletions are expected to be about 6% in summer and about 10% in winter. In addition, there is the possibility of incurring widespread losses in the Arctic. Enhanced levels of stratospheric sulphate aerosols from natural (e.g., Mt. Pinatubo) or anthropogenic sources could possibly lead to even greater ozone losses by increasing the catalytic efficiency of chlorine- and bromine-containing chemicals through heterogeneous chemical processes.

A reduction in the peak chlorine and bromine levels, and a hastening of the subsequent decline of these levels, hence reducing future levels of O_3 depletion, can be accomplished in a variety of ways, including an accelerated phaseout of controlled substances and limitations on currently uncontrolled halocarbons.

A1.7 Tropospheric Ozone Precursors: Carbon Monoxide, Non-Methane Hydrocarbons and Nitrogen Oxides

Tropospheric O_3 is predicted to increase with increasing emissions of nitrogen oxides (NO_x), and with increasing emissions of carbon monoxide (CO), CH_4 and non-methane hydrocarbons (NMHC) when the atmospheric abundance of NO_x is greater than 20-30 pptv. The magnitude of ozone changes are predicted to exhibit marked variations with latitude, altitude and season. Differences between model calculations of O_3 increases from NO_x emissions are large (factor of ~3), but moderate from CH_4 emissions (~50%). The differences in the predicted spatial and temporal distributions of O_3 changes are particularly large for NO_x emissions but again moderate for CH_4 emissions. It should also be noted that increases in CH_4, CO and NMHC emissions lead to reduced OH values, while increased NO_x emissions lead to enhanced OH levels. As a result of these opposing effects, the sign of future OH changes cannot be predicted reliably. Uncertainties in present and future tropospheric OH concentrations lead to corresponding uncertainties in the lifetimes of many tropospheric species, e.g., CH_4, HCFCs and HFCs.

A1.7.1 *Trends*

There is little new information on the trends of these tropospheric O_3 precursors. Because of their relatively short atmospheric lifetimes, coupled with inadequate monitoring networks, the determination of their long-term trends and the spatial and temporal variability of their atmospheric distribution is very difficult. The atmospheric abundance of CO appears to be increasing in the Northern Hemisphere at about 1% per year (Khalil and Rasmussen, 1990), but there is no evidence of a significant trend in the Southern Hemisphere (Khalil and Rasmussen, 1990; Brunke *et al.*, 1990). There is also no evidence for trends in the atmospheric concentrations of NO_x or NMHC, except for ethane in the Northern Hemisphere (0.9±0.3% per year: Ehhalt *et al.*, 1991).

A1.7.2 *Sources and Sinks*

While it is clear that CO, NMHC, and NO_x all have significant natural and anthropogenic sources (IPCC, 1990) their budgets remain uncertain. Recent data (Leleiveld and Crutzen, 1991) suggests that the oxidation of HCHO in the liquid phase does not produce CO, but CO_2 directly, therefore the estimates of CO production from hydrocarbon oxidation may need to be revised downward. Hameed and Dignon (1992) report that NO_x emissions have increased by about 30% over the period 1970-1986. Tables A1.8 and A1.9 show both natural and anthropogenic sources of CO and NO_x.

A1.8 Sulphur-Containing Gases

Sulphur-containing gases emitted into the atmosphere through natural and anthropogenic processes affect the Earth's radiative budget by being transformed into sulphate aerosol particles that: (i) scatter sunlight back to space, thereby reducing the radiation reaching the Earth's surface; (ii) possibly increase the number of cloud condensation nuclei, thereby potentially altering the physical characteristics of clouds; and (iii) affect atmospheric chemical composition, e.g., stratospheric O_3, by providing surfaces for heterogeneous chemical processes (see Section A2.6 for a fuller discussion). Sulphate aerosols are important in both the troposphere where they have lifetimes of days to a week or so, as well as in the stratosphere where they have lifetimes of several years.

A1.8.1 *Sources*

Spiro *et al.* (1991) have prepared a detailed global inventory of the geographic distribution (1° by 1°) of natural and anthropogenic sulphur emissions. The estimated emissions of sulphur dioxide (SO_2) from biomass burning and of reduced sulphur gases from soils and vegetation have been reduced (Andreae, 1990; Hao *et*

Table A1.8: *Estimated Sources and Sinks of Carbon Monoxide (Tg CO per year)*

	WMO (1985)	Seiler & Conrad (1987)	Khalil & Rasmussen (1990)	Crutzen & Zimmerman (1991)
Primary Sources				
• Fossil Fuel	440	640±200	400-1000	500
• Biomass Burning	640	1,000±600	335-1400	600
• Plants	-	75±25	50-200	-
• Oceans	20	100±90	20-80	-
Secondary Sources				
• NMHC oxidation	660	900±500	300-1400	600
• methane oxidation	600	600±300	400-1000	630
Sinks				
• OH reaction	900±700	2000±600	2200	2050
• Soil uptake	256	390±140	250	280
• Stratos. oxidation	-	110±30	100	-

Table A1.9: *Estimated Sources of Nitrogen Oxides (Tg N per year)*

Natural	
• Soils	5-20 (1)
• Lightning	2-20 (2)
• Transport from Stratosphere	~1
Anthropogenic	
• Fossil Fuel Combustion	24 (3)
• Biomass Burning	2.5-13 (4)
• Tropospheric Aircraft	0.6

(1) Dignon *et al.*, 1991;   (2) Atherton *et al.*, 1991;
(3) Hameed and Dignon, 1992;   (4) Dignon and Penner, 1991

Table A1.10: *Estimated Sources of Short-lived Sulphur Gases (Tg S per year)*

Anthropogenic emissions (mainly SO_2)	70-80
Biomass burning (SO_2)	0.8-2.5
Oceans (DMS)	10-50
Soils and plants (DMS and H_2S)	0.2-4
Volcanic emissions (mainly SO_2)	7-10

al., 1991; Bates *et al.*, 1991; Spiro *et al.*, 1991). The best estimate of the magnitude of annual global anthropogenic emissions of SO_2 is between 70Tg and 80Tg (Semb 1985; Hameed and Dignon, 1988; Langner and Rodhe, 1991; Spiro *et al.*, 1991). Recent estimates of the annual global emissions of dimethyl sulphide (DMS) from the oceans (10-50TgS) (Bates *et al.*, 1987; Bates *et al.*, 1991; Langner and Rodhe, 1991; Spiro *et al.*, 1991) are significantly lower than given by IPCC (1990). Table A1.10 summarizes known emissions of key short-lived sulphur gases.

A1.8.2 Atmospheric Gas-Particle Conversion

Oxidation of SO_2 to aerosol sulphate occurs in the gas phase and in cloud droplets (aqueous-phase oxidation). The former may generate new particles or the newly formed H_2SO_4 may add to existing particles increasing their mass but not their number. Aqueous-phase oxidation does not result in new particle formation but only adds to the mass of sulphate. The dissolved sulphate may remain in the atmosphere as an aerosol particle upon cloud droplet evaporation (this appears to be the more frequent situation) or may be removed from the atmosphere in precipitation. The rates of these atmospheric oxidation reactions depend on the concentrations of OH for the gas-phase reaction and of oxidants (hydrogen peroxide (H_2O_2) and O_3) for the aqueous-phase reactions. The rates of the aqueous-phase reactions increase with cloud liquid-water content, increase with decreasing temperature, as a consequence of temperature dependent gas solubilities, and for the O_3 reaction, decrease with decreasing cloud water pH. The decrease in pH as the reaction proceeds can lead to a self limitation of that reaction. While the extent of the H_2O_2 oxidation may be limited by the amount of H_2O_2 present, the amount of O_3 is rarely limiting to the extent of oxidation by that species. Detailed description of the rate, extent and spatial distribution of these reactions depends on knowledge of the concentrations of the reagent species and, in the case of the aqueous-phase reactions, of the pertinent cloud properties.

In addition to the mass concentration of the sulphate

aerosol it is necessary to have information about the size distribution of the aerosol particles, since this size distribution affects the radiative and cloud nucleating properties of the aerosol. The evolution of the size distribution of an aerosol (in clear air) is the resultant of new particle formation and coagulation and removal processes. These processes depend, in a complex and incompletely understood way, on the properties of the existing aerosol and the rate of generation of new condensable material. The size distribution is a strong function of relative humidity, shifting to larger sizes with increasing relative humidity (Charlson *et al.*, 1987; d'Almmeida *et al.*, 1991). The size distribution is also greatly influenced by cloud processes (Hoppel *et al.*, 1990; Hegg *et al.*, 1990).

A1.8.3 Transport and Distribution

The atmospheric lifetimes of SO_2, DMS and hydrogen sulphide (H_2S) are a few weeks at most, and their atmospheric distributions are largely controlled by the distributions of their sources. The mean residence time of sulphate aerosols formed by gas-particle conversion in the troposphere is about a week. Two consequences of the short lifetimes are that the resulting distribution of tropospheric aerosols is inhomogeneous, and that these gases are not significant contributors to the stratospheric sulphate layer. To assess the climatic impact of these aerosols it is necessary to know their spatial distribution in much more detail than is the case for the longer-lived greenhouse gases, (a) because the radiative forcing due to aerosols varies spatially, (b) because cloud microphysical processes are nonlinear in the concentration of aerosol particles, and (c) because cloud forcing is nonlinear in the concentration of cloud droplets.

A1.8.4 Removal

Removal of submicrometer aerosol particles contributing to the radiative effects occurs largely by the precipitation process (e.g., Slinn, 1983). These particles are the dominant particles on which cloud particles form (cloud condensation nuclei, CCN); once a cloud droplet (of diameter of a few up to 20 micrometres) is formed, it is much more susceptible to scavenging and removal in precipitation than is the original submicrometre particle. The fraction of aerosol particles incorporated in cloud droplets on cloud formation is the subject of active current research. Earlier work has yielded a fairly wide spread in this fractional incorporation, based in part on limitations of then-existing techniques and in part on the absence of a single definition of incorporation efficiency (ten Brink *et al.*, 1987). More recent work indicates a high fractional incorporation at low concentrations of aerosol particles decreasing as the aerosol particle loading increases (Gillani *et al.*, 1992). Model calculations of the efficiency of

incorporation of aerosol particles into cloud droplets and precipitation are highly sensitive to assumptions and approach (Jenson and Charlson, 1984; Flossmann *et al.*, 1985; Hanel, 1987; Ahr *et al.*, 1989; Alheit *et al.*, 1990).

A1.8.5 Stratospheric Aerosols

Volcanic injections of sulphur are a major contributor to the stratospheric aerosol layer. Krueger (1991) used Nimbus 7 TOMS data to estimate that the eruption of Mt. Pinatubo in the Philippines in 1991 added about 20 million tons of SO_2 directly to the stratosphere, about 50% more than the eruption of Mt. El Chichon. Anthropogenic emissions add to stratospheric sulphur and their magnitude needs to be evaluated. Carbonyl sulphide (COS), a significant source, is produced by the oxidation of carbon disulphide (CS_2) in the troposphere but its origins are anthropogenic. In addition, Hofmann (1990, 1991) has suggested that the abundance of stratospheric sulphate aerosols had increased during the last decade, possibly due to aircraft emissions of SO_2.

References

Ahr, M., A.I. Flossmann and H.R. Pruppacher, 1989: A comparison between two formulations for nucleation scavenging. *Beitr. Phys. Atmosphere*, **62**, 321-326.

Alheit, R.R., A.I. Flossmann and H.R. Pruppacher, 1990: A theoretical study of the wet removal of atmospheric pollutants - Part IV: The uptake and redistribution of aerosol particles through nucleation and impaction scavenging by growing cloud drops and ice particles. *J. Atmos. Sci.*, **47**, 870-887.

Andreae, M.O., 1990: Ocean-atmosphere interactions in the global biogeochemical sulphur cycle. *Mar. Chem.*, **30**, 1-29.

Atherton, C.S., J.E. Penner and J.J. Walton, 1991: The role of lightning in the tropospheric nitrogen budget: model investigations. *Geophys. Res. Lett.* (To be submitted).

Barns, D.W. and J.A. Edmonds, 1990: An evaluation of the relationship between the production and use of energy and atmospheric methane emissions, US Department of Energy, DOE/NBB-0088P.

Bates, T.S., J.D. Cline, R.H. Gammon and S.R. Kelly-Hansen, 1987: Regional seasonal variations in the flux of oceanic dimethylsulphide to the atmosphere. *Geophys. Res.*, **92**, 2930-2938.

Bates, T.S., B.K. Lamb and A. Guenther, 1991: Sulphur emissions to the atmosphere from natural sources. *J. Atmos. Chem.* (Submitted).

Bazzaz, F.A., 1990: The response of natural ecosystems to the rising global CO_2 levels. *Ann. Rev. Ecol. Syst.*, **21**, 67-196.

Blake, D.R. and F.S. Rowland, 1988: Continuing worldwide increase in tropospheric methane, 1978-1987. *Science*, **239**, 1129-1131.

Bouwman, A.F., 1990: Exchange of greenhouse gases between terrestrial ecosystems and the atmosphere. In: *Soil and the greenhouse effect*. A.F. Bouwman (Ed.). Wiley, pp62-125.

Bowden, R.D., P.A. Steudler, J.M. Melillo and J.D. Durham, 1990: Annual nitrous oxide fluxes from temperate forest soils in the northeastern United States. *J. Geophys. Res.*, **95**, 13,997-14,005.

Boyer, C.M., J.R. Kelafant, V.A. Kuuskraa and K.C. Manger, 1990: Methane emissions from coal mining: Issues and opportunities for reduction. US Environmental Protection Agency, EPA.400/9-90/008.

Brown, S. and A.E. Lugo, 1984: Biomass of tropical forests: A new estimate based on forest volumes. *Science*, **223**, 1290-1293.

Brown, S. and A.E. Lugo, 1991: Biomass of tropical forests of south and southeast Asia. *Can. J. For. Res.*, **21**, 111-117.

Brunke, E., H. Scheele and W. Seiler, 1990: Trends of tropospheric CO, N_2O and CH_4 as observed at Cape Point, South Africa. *Atmos. Environ.*, **24A**, 585-595.

Casada, M.E. and L.M. Safley, 1990: Global methane emissions from livestock and poultry manure. Report to the Global Change Division, US Environmental Protection Agency, Washington, D.C.

Charlson, R.J., J.E. Lovelock, M.O. Andreae and S.G. Warren, 1987: Oceanic phytoplankton, atmospheric sulfur, cloud albedo and climate. *Nature*, **326**, 655-661.

Cicerone, R.J. and J.D. Shetter, 1981: Sources of atmospheric methane: Measurements in rice paddies and a discussion. *J. Geophys. Res.*, **86**, 7203-7209.

Cicerone, R.J., J.D. Shetter and C.C. Delwiche, 1983: Seasonal variation of methane flux from a California rice paddy. *J. Geophys. Res.*, **88**, 11,022-11,024.

Cofer III, W.R., J.S. Levine, E.L. Winstead and B.J. Stocks, 1991: New estimates of nitrous oxide emission from biomass burning. *Nature*, **349**, 689-691.

CONCAWE, 1991: An EC-12/world inventory of greenhouse gas emissions from fossil fuel use. CONCAWE, Report No. 91/54, Brussels.

Crutzen, P.J. and P.H. Zimmermann, 1991: The changing photochemistry of the troposphere. *Tellus*, **43**, 136-151.

Crutzen, P.J., 1991: Methane's sinks and sources. *Nature*, **350**, 380-381.

d'Almmeida, G.A., P. Koepke and E.P. Shettle, 1991: *Atmospheric aerosols: Global climatology and radiative characteristics*. Deepak Publishing, Hampton, Virginia

De Soete, G., 1989: Updated evaluation of nitrous oxide emissions from industrial fossil fuel combustion. Report to the European Atomic Energy Community. Institut Francais du Petrole, Ref. 37-559.

Debo, L., Z. Jiwu, L. Weixing, Y. Fei, 1992: Monitoring and controlling of CH_4 emissions from the rice paddy field near Nanjing, China. In: *Climate–Biosphere Interactions: Biogenic Emissions and Environmental Effects of Climate Change*. R.G. Zepp (Ed.), Wiley, New York. (In press).

Denmead, O.T. and J.R. Freney, 1990: Micrometeorological measurements of methane emissions from flooded rice. Poster paper, SCOPE/IGBP Workshop on Trace Gas Exchange in a Global Perspective, Feb. 19-23, Sigtuna, Sweden.

Dignon, J. and J.E. Penner, 1991: Biomass burning: A source of nitrogen oxides in the atmosphere. In: *Global Biomass Burning: Atmospheric, Climatic and Biospheric Implications*. J.S. Levine (Ed.), MIT Press.

Dignon, J., J.E. Penner, C.S. Atherton and J.J. Walton, 1991: Atmospheric reactive nitrogen: a model study of natural and anthropogenic sources and the role of microbial soil emissions. Presented at CHEMRAWN VII, Baltimore, Maryland, December 2-6.

Drake, B.G. and P.W. Leadley, 1991: Canopy photosynthesis of crops and native communities exposed to long-term elevated CO_2. *Plant, Cell and Environment*, **14**, 853-860.

Ehhalt, D.H., U. Schmidt, R. Zander, P. Demoulin and C.P. Rinsland, 1991: Seasonal cycle and secular trend of the total and tropospheric column abundance of ethane above the Jungfraujoch. *J. Geophys. Res.*, **96**, 4985-4994.

Eichner, M.J., 1990: Nitrous oxide emissions from fertilized soils: Summary of available data. *J. Environ. Qual.*, **19**, 272-280.

Enting, I.G. and J.V. Mansbridge, 1991: Latitudinal distribution of sources and sinks of CO_2: results of an inversion study. *Tellus*, **43B**, 156-170.

FAO, 1990: Interim report on forest resources assessment 1990 project. Committee on Forestry, Tenth Session, FAO, Rome.

FAO, 1991: Second interim report on the state of tropical forests. 10th World Forestry Congress, Paris, France.

FAO/UNEP, 1981: Tropical Forest Resources Assessment Project. FAO, Rome.

Fearnside, P.M., A.T. Tardin and L.G. Meira Filho, 1990: Deforestation rate in Brazilian Amazonia. National Secretariat of Science and Technology, Brazilia, 8pp.

Flossmann, A.I., W.D. Hall and H.R. Pruppacher, 1985: A theoretical study of the wet removal of atmospheric pollutants - Part I: The redistribution of aerosol particles captured through nucleation and impaction scavenging by growing cloud drops. *J. Atmos. Sci.*, **42**, 583-606.

Fung, I., J. John, J. Lerner, E. Matthews, M. Prather, L.P. Steele and P.J. Fraser, 1991: Three-dimensional model synthesis of the global methane cycle. *J. Geophys. Res.*, **D7**, 13,033-13,065.

Gillani, N.V., P.H. Daum, S.E. Schwartz, W.R. Leaitch, J.W. Strapp and G.A. Isaac, 1992: Fractional activation of accumulation-mode particles in warm continental stratiform clouds. In: *Precipitation scavenging and atmosphere-surface exchange*. S.E. Schwartz and W.G.N. Slinn (Coordinators). Hemisphere, Washington D.C.

Grulke, N.E., G.H. Reichers, W.C. Oechel, U. Hjelm and C. Jaeger, 1990: Carbon balance in tussock tundra under ambient and elevated atmospheric CO_2. *Oecologia*, **83**, 485-494.

Hameed, S. and J. Dignon, 1988: Changes in the geographical distributions of global emissions of NO_x and SO_x from fossil fuel combustion between 1966 and 1980. *Atmos. Environ.*, **22**, 441-449.

Hameed, S. and J. Dignon, 1992: Global emissions of nitrogen and sulfur oxides in fossil fuel combustion 1970-1986. *J. of Air and Waste Management Assoc.*, **42**.

Hanel, G., 1987: The role of aerosol properties during the condensational growth of cloud: A reinvestigation of numerics and microphysics. *Beitr. Phys. Atmosphere*, **60**, 321-339.

Hao, W.M., M.H. Lie and P.J. Crutzen, 1991: Estimates of annual and regional releases of CO_2 and other trace gases to the atmosphere from fires in the tropics based on the FAO statistics for the period 1975-1980. In: *Fire In the Tropical Biota.* J.G. Goldammer (Ed.). Springer Verlag, Berlin, 440-462.

Hargraves, A.J., 1990: Coal seam gas and the atmosphere. In: *Greenhouse and Energy.* D.J. Swaine (Ed.). Melbourne, Australia, 147-156.

Harriss, R. 1991: Personal communication.

Hegg, D.A., L.F. Radke and P.V. Hobbs, 1990: Particle production associated with marine clouds. *J. Geophys. Res.,* **95**, 13,917-13,926.

Heimann, M., 1991: Modelling the global carbon cycle. In: *First Demetra Meeting on Climate Variability and Global Change.* Unpublished manuscript.

Heimann, M. and C.D. Keeling, 1989: A three-dimensional model of atmospheric CO_2 transport based on observed winds: 2. Model description and simulated tracer experiments. In: Aspects of climate variability in the Pacific and the Western Americas. D.H. Peterson (Ed.). *Geophysical Monograph,* **55**, AGU, Washington (USA), 237-275.

Heimann, M., C.D. Keeling and C.J. Tucker, 1989: A three-dimensional model of atmospheric CO_2 transport based on observed winds: 3. Seasonal cycle and synoptic time-scale variations. In: Aspects of climate variability in the Pacific and the Western Americas. D.H. Peterson (Ed.). *Geophysical Monograph,* **55**, AGU, Washington (USA), 277-303.

Hofmann, D.J., 1990: Increase in the stratospheric background sulphuric acid aerosol mass in the past 10 years. *Science,* **248**, 996-1000.

Hofmann, D.J., 1991: Aircraft sulphur emissions. *Nature,* **349**, 659.

Holzapfel-Pschorn, A. and W. Seiler, 1986: Methane emission during a cultivation period from an Italian rice paddy. *J. Geophys. Res.,* **91**, 11,803-11,814.

Hoppel, W.A., J.W. Fitzgerald, G.M. Frick, R.E. Larson and E.J. Mack, 1990: Aerosol size distributions and optical properties found in the marine boundary layer over the Atlantic Ocean. *J. Geophys. Res.,* **95**, 3659-3686.

Houghton, R.S., 1991: Tropical deforestation and atmospheric carbon dioxide. *Climatic Change.* (In press).

INPE, 1991: Unpublished data. Instituto Nacional de Pesquisas Espaciais, São Paulo, Brazil.

IPCC, 1990: *Climate Change, The IPCC Scientific Assessment.* WMO/UNEP. J.T. Houghton, G.J. Jenkins and J.J. Ephraums (Eds.). Cambridge University Press, Cambridge (UK), pp365.

Jenson, J.B. and R.J. Charlson, 1984: On the efficiency of nucleation scavenging. *Tellus,* **36B**, 367-375.

Johnson, D.W., C.G. Kilsby, D.S. McKenna, R.W. Saunders, G.J. Jenkins, F.B. Smith and J.S. Foot, 1991: Airborne observations of the physical and chemical characteristics of the Kuwait oil smoke plume. *Nature,* **353**, 617-621.

Keeling, C.D., R.B. Bacastow, A.F. Carter, S.C. Piper, T.P. Whorf, M. Heimann, W.G. Mook and H. Roeloffzen, 1989a: A three-dimensional model of atmospheric CO_2 transport based on observed winds: 1. Analysis of observational data. In: Aspects of climate variability in the Pacific and the Western Americas. D.H. Peterson (Ed.). *Geophysical Monograph,* **55**, AGU, Washington (USA), 165-236.

Keeling, C.D., S.C. Piper and M. Heimann, 1989b: A three-dimensional model of atmospheric CO_2 transport based on observed winds: 4. Mean annual gradients and interannual variations. In: Aspects of climate variability in the Pacific and the Western Americas. D.H. Peterson (Ed.). *Geophysical Monograph,* **55**, AGU, Washington (USA), 305-363.

Keller, M., 1992: Controls on the soil-atmosphere fluxes of nitrous oxide and methane: Effects of tropical deforestation. In: *Climate-Biosphere Interactions: Biogenic Emissions and Environmental Effects of Climate Change.* R.G. Zepp (Ed.). Wiley, New York. (In press).

Keller, M., M.E. Mitre and R.F. Stallard, 1990: Consumption of atmospheric methane in soils of Central Panama: Effects of agricultural development. *Global Biogeochem. Cycles,* **4**, 21-27.

Kempe, S. and K. Pegler, 1991: Sinks and sources of CO_2 in coastal seas: the North Sea. *Tellus,* **43B**, 224-235.

Khalil, M.A.K. and R.A. Rasmussen, 1990: The global cycle of carbon monoxide: Trends and mass balance. *Chemosphere, 20, 227-242.*

Khalil, M.A.K. and R.A. Rasmussen, 1991: Methane emissions from rice fields in China. *Environ. Sci. Technol.,* **25**, 979-981.

Khalil, M.A.K., R.A. Rasmussen, J.R.J. French and J.A. Holt, 1990: The influence of termites on atmospheric trace gases; CH_4, CO_2, $CHCl_3$, N_2O, CO, H_2 and light hydrocarbons. *J. Geophys. Res.,* **95D**, 3619-3634.

Krueger, A., 1991: Unpublished TOMS satellite data.

Langner, J. and H. Rodhe, 1991: A global three-dimensional model of the tropospheric sulphur cycle. *J. Atmos. Chem.,* **13**, 225-263.

Lasaga, A.C. and G.V. Gibbs, 1991: Ab initio studies of the kinetic isotope effect of the CH_4 + OH atmospheric reaction. *Geophys. Res. Lett.,* **18**, 1217-1220.

Leleiveld, J. and P.J. Crutzen, 1991: The role of clouds in tropospheric photochemistry. *J. Atmos. Chem.,* **12**, 229-267.

Linak, W.P., J.A. McSorley, R.E. Hall, J.V. Ryan, R.K. Srivastava, J.O.L. Wendt and J.B. Mereb, 1990: Nitrous oxide emissions from fossil fuel combustion. *J. Geophys. Res.,* **95**, 7533-7541.

Lobert, J.M., D.H. Scharffe, W.M., Hao and P.J. Crutzen, 1990: Importance of biomass burning in the atmospheric budgets of nitrogen-containing gases. *Nature,* **346**, 552-556.

Lugo, A.E. and S. Brown, 1992: Tropical forests as sinks of atmospheric carbon. Unpublished manuscript.

Manning, M.R., D.C. Lowe, W. Melhuish, R. Spaarks, G. Wallace, C.A.M. Brenninkmeijer and R.C. McGill, 1990: The use of radiocarbons measurements in atmospheric studies. *Radiocarbons,* **32**, 37-58.

Marland, G. and T. Boden, 1991: Unpublished data.

Matson, P.A., P.M. Vitousek, G.P. Livingston and N.A. Swanberg, 1990: Sources of variation in nitrous oxide flux from Amazonian ecosystems. *J. Geophys. Res.,* **95**, 16,789-16,798.

McCulloch, A., 1991: Personal communication.

Mitra, A.P., 1992: Global change: Greenhouse gas emissions in India. Scientific Report #2. Prepared for the Council of Scientific and Industrial Research and Ministry of Environment and Forests.

Montreal Protocol on Substances that Deplete the Ozone Layer, Final Act, 1987: United Nations Environment Programme, Na.87-6106. Available from UNEP Ozone Secretariat, Nairobi, Kenya. (1990 London Amendments to Protocol also available).

Mosier, A., D. Schimel, D. Valentine, K. Bronson and W. Parton, 1991: Methane and nitrous oxide fluxes in native, fertilized and cultivated grasslands. *Nature*, **350**, 330-332.

Murphy, P.P., R.A. Feely, R.H. Gammon, D.E. Harrison, K.C. Kelly and L.S. Waterman, 1991: Assessment of the Air-Sea Exchange of CO_2 in the South Pacific during Austral Autumn. *J. Geophys. Res.*, **96**, 20,455-20,465.

Myers, N., 1980: Conversion of tropical moist forest. US National Academy of Sciences, Washington, DC.

Myers, N., 1989: Deforestation rates in tropical forests and their climatic implications. Friends of the Earth, London, UK.

Oechel, W.C. and B.R. Strain, 1985: Native species responses to increased carbon dioxide concentration. In: *Direct Effects of Increasing Carbon Dioxide on Vegetation*. B.R. Strain and J.D. Cure (Eds.). US Department of Energy, NTIS, Springfield, Virginia, pp117-154.

Okken, P.A. and T. Kran, 1989: CH_4/CO emissions from fossil fuels-global potential. Prepared for IEA/ETSAP Workshop, Paris, June 1989. Petten, The Netherlands: Energieonderzoek Centrum ESC.

Parashar, D.C., J. Rai, P.K. Gupta and N. Singh, 1991: Parameters affecting methane emission from paddy fields. *Indian Journal of Radio and Space Physics*, **20**, 12-17.

Parton, W.J., D.S. Schimel, C.V. Cole and D.S. Ojima, 1987: Analysis of factors controlling soil organic matter levels in Great Plains grasslands. *Soil Sci. Soc. Amer. Journal* , **51**, 1173-1179.

Prinn, R., D. Cunnold, R. Rasmussen, P. Simmonds, F. Alyea, A. Crawford, P. Fraser and R. Rosen, 1990: Atmospheric emissions and trends of nitrous oxide deduced from 10 years of ALE-GAGE data. *J. Geophys. Res.,* **95**, No. D11, p18,369-18,385.

Prinn, R., F. Alyea, D. Cunnold, P. Fraser and P. Simmonds, 1991: Unpublished data from the ALE-GAGE network.

Prinn, R., D. Cunnold, P. Simmonds, F. Alyea, R. Boldi, A. Crawford, P. Fraser, D. Gutzler, D. Hartley, R. Rosen and R. Rasmussen, 1992: Global average concentration and trend for hydroxyl radicals deduced from ALE/GAGE trichloroethane (methyl chloroform) data for 1978-1990. *J. Geophys. Res.* (Submitted).

Prudhomme, T.I., W.C. Oechel, S.J. Hastings and W.T. Lawrence, 1984: Net ecosystem gas exchange at ambient and elevated carbon dioxide concentrations in tussock tundra at Toolik Lake, Alaska: an evaluation of methods and initial results. In: *The Potential Effects of Carbon Dioxide-Induced Climatic Changes in Alaska: Proceedings of a Conference School of Agricultural and Land Resources Management*. J.H. McBeath (Ed.). University of Alaska, Fairbanks, Alaska.

Quay, P.D., S.L. King, J. Stutsman, D.O. Wilbur, L. P. Steele, I. Fung, R.H. Gammon, T.A. Brown, G.W. Farwell, P.M. Grootes and F.H. Schmidt, 1991: Carbon isotopic composition of atmospheric CH_4: Fossil and biomass burning source strengths. *Global Biogeochem. Cycles*, **5**, 25-47.

Quay, P.D., B. Tilbrook and C.S. Wong, 1992: Oceanic uptake of fossil fuel CO_2: Carbon-13 evidence. Unpublished manuscript.

Raich, J.W., E.B. Rastetter, J.M. Melillo, D.W. Kicklighter, P.A. Steudler, B.J. Peterson, A.L. Grace, B. Moore III and C.J. Vorosmarty, 1991: Potential net primary production in South America: application of a global model. *Ecol. Appl.*, **1**, 399-429.

Rouland, C., A. Brauman, M. Labat, M. Lepage and J.C. Mennaut, 1991: Poster paper presented at the NATO Advanced Research Workshop. The Atmospheric Methane Cycle: Sources, Sinks, Distribution and Role in Global Change. Portland, Oregon, 6-11 October 1991.

Sanhueza, E., W.M. Hao, D. Scharffe, L. Donoso and P.J. Crutzen, 1990: N_2O and NO emissions from soils of the Northern part of the Guayana Shield, Venezuela. *J. Geophys. Res.*, **95**, 22,481-22,488.

Sass, R.L., F.M. Fisher, P.A. Narcombe and F.T. Turner, 1990: Methane production and emission in Texas rice field. *Global Biogeochem. Cycles*, **4**, 47-68.

Scharffe, D., W.M. Hao, L. Donoso, P.J. Crutzen and E. Sanhueza, 1990: Soil fluxes and atmospheric concentration of CO and CH_4 in the Northern part of the Guayana Shield, Venezuela. *J. Geophys. Res.*, **95**, 22,475-22,480.

Schutz, H., A. Holzapfel-Pschorn, R. Conrad, H. Rennenberg and W. Seiler, 1989: A 3-year continuous record on the influence of daytime, season and fertilizer treatment on methane emission rates from an Italian rice paddy. *J. Geophys. Res.*, **94**, 16,405-16,416.

Seiler, W. and R. Conrad, 1987: Contribution of tropical ecosystems to the global budget of trace gases, especially CH_4, H_2, CO and N_2O. In: *The Geophysiology of Amazonia*. R. Dickenson (Ed.). John Wiley, New York. 133-160.

Seiler, W., A. Holzapfel-Pschorn, R. Conrad and D. Scharffe, 1984: Methane emission from rice paddies. *J. Atmos. Chem.*, **1**, 241-268.

Semb, A. 1985: *Circumpolar SO_2 emission survey*. Norwegian Institute for Research, Lillestrom, Norway. NILU OR 69/85.

Setzer, A.W. and M.C. Pereira, 1991: Amazonia biomass burnings in 1987 and an estimate of their tropospheric emissions. *Ambio*, **20**, 19-22.

Slinn, W.G.N., 1983: Air to sea transfer of particles. In: *Air-Sea Exchange of Gases and Particles*. P.S. Liss and W.G.N. Slinn (Eds.). D. Reidel, Dordrecht. pp299-405.

Spiro, P.A., D.J. Jacob and J.A. Logan, 1991: Global inventory of sulphur emissions with $1° \times 1°$ resolution. *J. Geophys. Res.* (Submitted).

Steele, L.P., E.J. Dlugokencky, P.M. Lang, P.P. Tans, R.C. Martin and K.A. Masarie, 1992: Slowing down of the global accumulation of atmospheric methane during the 1980s. *Nature*. (Submitted).

Stevens, C.M., 1988: Atmospheric methane. *Chem. Geology*, **71**, 11-21.

Stolarski, R.S., P. Bloomfield, R.D. McPeters and J.R. Herman, 1991: Total ozone trends deduced from Nimbus 7 TOMS data. *Geophys. Res. Lett.*, **18**, 1015-1018.

Stordal, F. and G. Myhre, 1991: *Greenhouse Effect and Greenhouse Warming Potential for SF_6*. Norwegian Institute for Air Research. NILU OR:74/91, 0-91065.

Tans, P.P., I.Y. Fung and T. Takahashi, 1990: Observational constraints on the global atmospheric carbon dioxide budget. *Science*, **247**, 1431-1438.

ten Brink, H., S. Schwartz and P.H. Daum, 1987: Efficient scavenging of aerosol sulfate by liquid-water clouds. *Atmos. Environ.*, **21**, 2035-2052.

Thiemens, M.H. and W.C. Trogler, 1991: Nylon production: An unknown source of atmospheric nitrous oxide. *Science*, **251**, 932-934.

Thornley, J.H.M., D. Fowler and M.G.R. Cannell, 1991: Terrestrial carbon storage resulting from CO_2 and nitrogen fertilization in temperate grasslands. *Plant, Cell and Environment*, **14**, 1007-1011.

Tissue, D.T. and W.C. Oechel, 1987: Response of *Eriophorum vaginatum* to elevated CO_2 and temperature in the Alaskan tussock tundra. *Ecology*, **68**, 401-410.

UNEP, 1991: Montreal Protocol 1991 Assessment Report of the Technology and Economic Assessment Panel, December 1991.

Vaghjiani, G.L. and A.R. Ravishankara, 1991: New measurement of the rate coefficient for the reaction of OH with methane. *Nature*, **350**, 406-409.

Wang, M.-X., D. Aiguo, S. Renxing, H. Schutz, H. Rennenberg, W. Seiler and W. Haibao, 1989: CH_4 emission from a Chinese rice paddy field. *Acta Meteorologica Sinica*, **4**, 265-275.

Watson, A.J., C. Robinson, J.E. Robinson, P.J. le B. Williams and M.J.R. Fasham, 1991: Spatial variability in the sink for atmospheric carbon dioxide in the North Atlantic. *Nature*, **350**, 50-53.

Whalen, M. and W. Reeburg, 1990: Consumption of atmospheric methane by tundra soils. *Nature*, **346**, 160-162.

Whalen, M., N. Takata, R. Henry, B. Deck, J. Zeglen, J.S. Vogel, J. Southon, A. Shemesh, R. Fairbanks and W. Broecker, 1989: Carbon-14 in methane sources and in atmospheric methane: The contribution from fossil carbon. *Science*, **245**, 286-290.

Whalen, M., W. Reeburg and K. Sandbeck, 1990: Rapid methane oxidation in a landfill cover soil. *Appl. Environ. Microbiol.*, **56**, 3045-3411.

WMO, 1985: *Atmospheric Ozone 1985*. Chapter 15, WMO Global Ozone Research and Monitoring Project, Report No. 16, Geneva.

WMO, 1989: *Report of the International Ozone Trends Panel 1988*. WMO Global Ozone Research and Monitoring Project, Report No. 18, Geneva.

WMO, 1992: *Scientific Assessment of Ozone Depletion:1991*. WMO/UNEP, WMO Global Ozone Research and Monitoring Project, Report No. 25, Geneva

WRI, 1990: *World Resources 1990-1991*. World Resources Institute and United Nations Development Programme. Oxford University Press, New York.

Yagi, K. and K. Minami, 1990: Effect of organic matter application on methane emission from some Japanese paddy field. *Soil Sci. Plant Nutr.*, **36**, 599-610.

Yagi, K., 1991: Unpublished data.

Yokoyama, T., S. Nishinomiya and H. Matsuda, 1991: N_2O emission from fossil fuel fired power plants. *Environ. Sci. Technol.*, **25**, 347-348.

Zongliang, C., S. Kesheng and W. Bujun, 1992: Methane emissions from different cultivation practices of rice paddies in Beijing. In: *Climate-Biosphere Interactions: Biogenic Emissions and Environmental Effects of Climate Change*. R.G. Zepp (Ed.). Wiley, New York. (In press).

A2

Radiative Forcing of Climate

I.S.A. ISAKSEN, V. RAMASWAMY, H. RODHE, T.M.L. WIGLEY

Contributors:
R. Charlson; I. Karol; J. Lelieveld; C.B. Leovy; S.E. Schwartz; K.P. Shine; R.T. Watson; D.J. Wuebbles; B.A. Callander

CONTENTS

EXECUTIVE SUMMARY

Since the 1990 IPCC Scientific Assessment (IPCC, 1990), there have been significant advances in our understanding of the impact of ozone depletion and sulphate aerosols on radiative forcing and of the limitations of the concept of the Global Warming Potential (GWP).

Radiative Forcing due to Changes in Stratospheric Ozone:
Observed global depletions of ozone (O_3) in the lower stratosphere have been used to calculate changes in the radiative balance of the atmosphere. Although the results are sensitive to atmospheric adjustments, and no General Circulation Model (GCM) studies of the implications of the O_3 changes on surface temperature have been performed, the radiative balance calculations indicate that the O_3 reductions observed during the 1980s have caused reductions in the radiative forcing of the surface-troposphere system at mid- and high latitudes. This reduction in radiative forcing resulting from O_3 depletion could, averaged on a global scale and over the last decade, be approximately equal in magnitude and opposite in sign to the enhanced radiative forcing due to increased chlorofluorocarbons (CFCs) during the same time period.

Radiative Forcing due to Changes in Tropospheric Ozone:
While there are consistent observations of an increase in tropospheric ozone (up to 10% per decade) at a limited number of locations in Europe, there is not an adequate global set of observations to quantify the magnitude of the increase in radiative forcing. However, it has been calculated that a 10% uniform global increase in tropospheric ozone would increase radiative forcing significantly compared to forcing by other greenhouse gases.

Radiative Effects of Sulphur Emissions:
Emissions of sulphur compounds from anthropogenic sources lead to the presence of sulphate aerosols which reflect solar radiation. This is likely to have a cooling influence on the Northern Hemisphere. For clear-sky conditions alone, the cooling caused by current rates of emissions has been estimated to be up to $1 Wm^{-2}$ averaged over the Northern Hemisphere, a value which should be compared with the estimate of $2.5 Wm^{-2}$ for the heating due to anthropogenic greenhouse gas emissions up to the present. The non-uniform distribution of anthropogenic sulphur sources coupled with the relatively short atmospheric residence time of sulphur compounds produce large regional variations in their effects. In addition, sulphate aerosols may affect the radiation budget through changes in cloud optical properties.

Global Warming Potentials:
Gases can exert a radiative forcing both directly and indirectly: direct forcing occurs when the gas itself is a greenhouse gas; indirect forcing occurs when chemical transformation of the original gas produces a gas or gases which themselves are greenhouse gases. The concept of the Global Warming Potential (GWP) has been developed for policymakers as a measure of the possible warming effect on the surface-troposphere system arising from the emission of each gas relative to carbon dioxide (CO_2). The indices are calculated for the contemporary atmosphere and do not take into account possible changes in chemical composition of the atmosphere. Changes in radiative forcing due to CO_2 are highly non-linear with respect to changes in atmospheric CO_2 concentrations. Hence, as CO_2 levels increase from present values, the GWPs of the non-CO_2 gases would be higher than those evaluated here. For the concept to be most useful, both the direct and indirect components of the GWP need to be quantified.

Direct Global Warming Potentials:
The direct components of the GWPs have been recalculated, taking into account revised estimated lifetimes, for a set of time horizons ranging from 20 to 500 years, with CO_2 as the reference gas. The same ocean-atmosphere carbon cycle model as used in IPCC (1990) has been used to relate CO_2 emission to concentrations. While in most cases the values are similar to the previous IPCC (1990) values, the GWPs for some of the hydrochlorofluorocarbons (HCFCs) and hydrofluorocarbons (HFCs) have increased by 20 to 50% because of revised estimates of their lifetimes. The direct GWP of methane (CH_4) has been adjusted upward, correcting an error in the previous IPCC report. The carbon cycle model used in these calculations probably underestimates both the direct and indirect GWP values for all non-CO_2 gases. The magnitude of the bias depends on the atmospheric lifetime of the gas, and the GWP time horizon.

Indirect Global Warming Potentials:
Because of our incomplete understanding of chemical processes, most of the indirect GWPs reported in IPCC (1990) are likely to be in substantial error, and none of them can be recommended. Although we are not yet in a position to recommend revised

numerical values, we do know that the indirect GWP for methane is positive and could be comparable in magnitude to its direct value. In contrast, the indirect GWPs for chlorinated and brominated halocarbons are likely to be negative. The concept of a GWP for short-lived, heterogeneously distributed constituents, such as carbon monoxide (CO), non-methane hydrocarbons (NMHC), and nitrogen oxide (NO_x) may prove inapplicable, although these constituents will affect the radiative balance of the atmosphere through changes in tropospheric ozone and the hydroxyl radical (OH). Similarly, a GWP for sulphur dioxide (SO_2) is viewed to be inapplicable because of the non-uniform distribution of sulphate aerosols.

A2.1 Introduction

This section is an update of the discussions presented in Section 2 of the first Intergovernmental Panel on Climate Change Scientific Assessment (IPCC, 1990) concerning the radiatively and chemically important species in the atmosphere. The present update has five major objectives:

(i) to extend the discussion on Global Warming Potentials (GWPs), and to re-evaluate their numerical values in view of revisions to the lifetimes of the radiatively active species;

(ii) to characterize the radiative impacts of the observed stratospheric O_3 losses between 1979 and 1990;

(iii) to discuss the changes in the concentrations of radiatively active gases occurring through chemical processes, and their implications for radiative forcing;

(iv) to characterize the radiative forcing due to tropospheric and stratospheric sulphate aerosol concentrations;

(v) to extend the discussion on the radiative forcing due to solar irradiance variations.

Some of the scientific details regarding the new developments have already appeared in the 1991 Scientific Assessment of Ozone Depletion (WMO, 1992).

A2.2 Radiative Forcing

The radiative forcing due to a perturbation in the concentration of a species is defined (see WMO, 1986; IPCC, 1990; WMO, 1992) by the net radiative flux change induced at the tropopause, keeping the concentrations of all other species constant. The change in the radiative flux is determined using a one-dimensional radiative transfer model in which the surface and tropospheric temperatures are held fixed at some reference values, while the stratospheric temperatures are allowed to relax to a new equilibrium in response to the perturbation. The definition assumes that the stratosphere undergoes a purely radiative adjustment, i.e., there is no change in the dynamical heating of the stratosphere due to the perturbation (WMO, 1986).

The radiative forcing is interpreted as a gain (positive) or loss (negative) for the surface-troposphere system as a whole. The motivation for this concept arises from radiative-convective modelling exercises, where the change in the global mean surface temperature can be related simply to the net radiative flux change at the tropopause (WMO, 1986). This has led to the adoption of the radiative forcing of the surface-troposphere system as a simple and convenient basis for estimating the potential climatic effects of various trace species.

A2.3 The Global Warming Potential (GWP) Concept

The aim of the GWP index is to offer a simple characterization of the relative radiative effects of the well-mixed species. It was created in order to enable policymakers to evaluate options that affect the emissions of various greenhouse gases, by avoiding the need to make repeated, complex calculations. IPCC (1990) discussed the GWP concept in considerable detail and only the salient features are addressed here. However, as noted below, there are serious limitations associated with the calculation of GWPs that constrain their practical utility.

A2.3.1 Definition

The Global Warming Potential is a measure of the relative, globally-averaged warming effect arising from the emissions of a particular greenhouse gas.

- It is a *relative* measure in that it expresses the warming effect compared to that of a reference gas (or 'molecule').

- It is a *global* measure in that it is derived from the globally- and annually-averaged net radiative fluxes at the tropopause, and thus describes the effects on the whole surface-troposphere system.

- It is a *time-integrated* measure of warming over a specified time horizon, taking account of the change with time of the species concentration.

The GWP of a well-mixed gas was defined in IPCC (1990) as the time-integrated change in radiative forcing due to the instantaneous release of 1kg of a trace gas expressed relative to that from the release of 1kg of CO_2. Calculation of the GWP for a particular species requires specification of the following:

(i) the radiative forcing both of the reference gas and of the species, per unit mass or concentration change;

(ii) the time horizon over which the forcings have to be integrated;

(iii) the atmospheric lifetime both of the species and of the reference gas;

(iv) the pathway of chemical breakdown of the species and the extent to which it gives rise to other greenhouse species, e.g., O_3 production from CH_4, NO_x, CO and NMHCs;

(v) the present and future chemical state of the atmosphere, i.e., levels of the background concentrations of various species throughout the troposphere;

(vi) the present and future physical state of the atmosphere, i.e., values of meteorological variables throughout the troposphere (e.g., temperature profile, cloud properties).

Factors (iii) and (iv) are intimately related to (v) and (vi) and are the sources of greatest uncertainty in the calculation of GWPs - see Section A2.3.4 below.

It is possible to offer alternative definitions of GWP, for example based on sustained rather than pulse emissions (Wigley *et al.*, 1990). Such alternatives can lead to numerical values of GWP which are different from those under the present definition, but in general not so different as to alter the relative ranking of the important species.

A2.3.2 Reference Molecule

Given the conceptual framework of the GWP and its implications for policy-making, the choice of a reference molecule is dictated by the need to evaluate the results in terms of the dominant contributor in the greenhouse gas problem. IPCC (1990) therefore chose CO_2 as the reference gas for the determination of GWPs. Although another gas or surrogate would have a simpler atmospheric decay behaviour compared to CO_2 (e.g., CFCs; see Fisher *et al.*, 1990), the evaluation of GWPs presented here continues, after extensive review, to use CO_2 as the reference gas.

To avoid the need to use a single lifetime for CO_2, IPCC (1990) used a carbon cycle model to calculate the integrated radiative forcing for CO_2, specifically the ocean-atmosphere box diffusion model of Siegenthaler and Oeschger (1987; see also Siegenthaler, 1983) which assumed a net neutral biosphere. We adopt the same model for the direct GWP estimates in this assessment.

A2.3.3 Time Horizons for GWPs

Because greenhouse gases have a variety of removal mechanisms they have different residence times, or lifetimes, in the atmosphere. The calculated value of GWP thus depends on the integration period chosen. There is no single value of integration time for determining GWPs that is ideal over the range of uses of this concept, though the choice of a time-scale for integration in the GWP calculation need not be totally arbitrary (see IPCC (1990) and WMO (1992) for a discussion on the choice of time horizons). In this report, GWPs are calculated over time horizons of 20, 100 and 500 years (as employed in IPCC, 1990). It is believed that these three time horizons provide a practical range for policy applications.

A2.3.4 Limitations of Present GWPs

While the GWP, as defined in IPCC (1990), is a convenient and reasonably practical index for ranking the relative and cumulative impact of greenhouse gas emissions, it has the following limitations, some of which are very serious:

(a) the modelling of radiative transfer within the atmosphere contains uncertainties, as was pointed out in IPCC (1990);

(b) since the direct GWP is a measure of the global effect of a given greenhouse gas emission, it is most appropriate for well-mixed gases in the troposphere (e.g., CO_2, CH_4, nitrous oxide (N_2O) and halocarbons). The radiative forcing employed in the determination of GWPs does not purport to characterize the latitudinal and seasonal dependence of the change in the surface-troposphere radiative fluxes. Different well-mixed gases can yield different spatial patterns of radiative forcings (Wang *et al.*, 1991);

(c) the GWP definition used here considers only the surface-troposphere radiative forcing rather than a particular response (e.g., surface temperature) of the climate system. While the surface-troposphere radiative flux perturbations can be related to temperature changes at the surface in the context of one-dimensional radiative-convective models (WMO, 1986), such a general interpretation for the temperature response either in three-dimensional General Circulation Models or in the actual surface-atmosphere system must be approached with caution. Further, although the GWP of a well-mixed gas can be regarded as a first-order indicator of the potential global mean temperature change due to that gas relative to CO_2, it is inappropriate for predicting or interpreting regional climate responses;

(d) GWP values are sensitive to uncertainties regarding atmospheric residence times. Thus, revisions to GWP values should be expected as scientific understanding improves. Because CO_2 is used as the reference gas, any revision to the calculation of its integrated radiative forcing over time will change all GWP values. GWP results are also sensitive to the choice of carbon cycle model used to calculate the time-integrated radiative forcing for CO_2. In particular, because the Siegenthaler-Oeschger model has only an ocean CO_2 sink, it is likely to overestimate the concentration changes and to lead to an underestimate of both the direct and the indirect GWPs. The magnitude of this bias depends on the atmospheric lifetime of the gas, and the time horizon;

(e) as defined here, GWPs assume constant concentration backgrounds at current levels. The calculated GWPs depend on the assumed background level(s). The indices are calculated for the contemporary atmosphere and do not take into account possible changes in the chemical composition of the atmosphere. Changes in radiative forcing due to CO_2, CH_4 and N_2O concentration changes are non-linear with respect to these changes. The net effect of these non-linearities is such that, as CO_2 levels increase from present values, the GWPs of all non-CO_2 gases would become higher than those evaluated here (see WMO, 1992);

(f) for the GWP concept to be most useful, both the direct and the indirect components need to be quantified. However, accurate estimates of the indirect effects are more difficult to obtain than those for the direct effects for the following reasons:

    (i) there are uncertainties in the details of the chemical processes as well as in the spatial and temporal variations of species involved in such transformations. As shown later, there is fair confidence in the sign of some of the indirect effects; however, precise estimates are lacking. Because of our incomplete understanding of chemical processes, it is now recognized that the uncertainties in the indirect components of GWPs reported in IPCC (1990) are so large that their use can no longer be recommended;

    (ii) for gases that are not well-mixed (e.g., tropospheric ozone precursors), the GWP concept may not be meaningful;

    (iii) further, while the GWP concept thus far has been applied to gases with perturbations only in the longwave spectra, it may not adequately account for the seasonally and latitudinally varying radiative effects due to inhomogeneously distributed species with a significant interaction in the solar spectrum (e.g., aerosols).

In conclusion, given the above limitations, great care must be exercised in applying GWPs in the policy arena.

A2.3.5 Direct GWPs of Well-Mixed Trace Gases

New direct GWPs (i.e., ignoring any radiative effects due to the products of chemical transformation) of several well-mixed species have been determined. For CFC-13, CFC-14, CFC-116, $CHCl_3$ and CH_2Cl_2 the lifetimes and radiative forcings are as given by Ramanathan *et al.* (1985); for the other compounds radiative forcings are as before (Tables 2.2 - 2.6 of IPCC, 1990) but lifetimes have been updated according to WMO (1992). Note that the lifetime of methane is here assumed to be 10.5 years, which accounts for a sink mechanism in the soil (see IPCC, 1990).

These new GWPs are listed in Table A2.1 for the three time horizons mentioned above. Changes in the lifetime and variations of radiative forcing with concentration change are neglected. Most of the new direct GWPs computed are generally within 20% of the values appearing in Table 2.8 of the IPCC (1990) study; the difference is entirely due to the differences in assumed lifetimes (WMO, 1992). HFC-125, HCFC-141b and HFC-143a, all have increases in GWPs exceeding 20% for the 100- and 500-year time horizons, while CF_3Br (Halon 1301) has a decrease of more than 20% for the 500-year time horizon; again, these differences are a manifestation

of the changes in the lifetimes. The direct GWPs for CH_4 here are substantially higher than those that can be inferred from IPCC (1990) owing to a typographical error appearing in that report. Note that CO, NMHC and NO_x have a negligible direct GWP component.

A2.3.6 Indirect GWPs of Well-Mixed Trace Gases

Because of our incomplete understanding of chemical processes, and their latitudinal and temporal dependence, it is not possible to quantify accurately the indirect GWPs. As noted above, the indirect GWPs reported in IPCC (1990) are likely to be in error and should not be used. In particular, the value for NO_x was probably overestimated by a substantial amount (Johnson *et al.*, 1992). Although we are not yet in a position to calculate new indirect GWPs, we can estimate the sign most likely for some compounds based on current understanding (see Table A2.1). For example, the indirect GWP for methane is positive and could be comparable in magnitude to the direct value. Because the weight of evidence suggests that halocarbons are largely responsible for the observed global stratospheric ozone loss over the past decade (WMO, 1992), chlorine- and bromine-containing compounds are likely to have negative indirect values. Although CO_2 is not itself involved in chemical reactions affecting the concentrations of radiatively active species, it could affect chemical processes through its influences on the atmospheric thermal structure. Other compounds such as CO, NMHC and NO_x indirectly affect the radiative balance of the atmosphere through changes in tropospheric ozone and OH (see Section A2.5). For such short-lived gases, however, the GWP concept may have little practical applicability.

A2.4 Radiative Forcing due to Stratospheric Ozone

A2.4.1 Lower Stratospheric Losses

There have been statistically significant losses of ozone in the middle and high latitudes between 1979 and 1990 which have important ramifications for radiative forcing (for details, the reader is referred to WMO, 1992). The TOMS satellite data (Stolarski *et al.*, 1991) point to a reduction in the column ozone, while the SAGE satellite data (McCormick *et al.*, 1992) and the ozonesonde data (Staehelin and Schmid, 1991) indicate that these losses have occurred mainly below 25km in the lower stratosphere. The weight of evidence suggests that these losses, both in the polar and the middle latitudes, are due in large part to the anthropogenic emissions of CFCs, as well as to other chlorine- and bromine-containing compounds (WMO, 1992).

These changes in ozone substantially perturb both solar and longwave radiation (WMO, 1986 and 1992). While the solar effects due to ozone losses are determined solely by

Table A2.1: *Numerical estimates of the "Direct" GWP and the sign of the "Indirect" effect of several gases. The lifetimes of various non-CO₂ species follow WMO (1992) and Ramanathan et al. (1985). The carbon cycle model employed follows IPCC (1990). The 20, 100 and 500-year time horizons denote the time elapsed after a pulse release of the gas in consideration. The radiative forcing values follow IPCC (1990) and Ramanathan et al. (1985). The following important points should be noted regarding the entries in the Table:*

1. *The lifetimes of the various species are not as precisely known as the Table indicates; they are used for the GWP calculations primarily to ensure consistency with the lifetime values in the references cited.*
2. *This assessment does not calculate the "Total" (direct + indirect) GWP as did IPCC (1990). Note that CO, NMHC and NO$_x$ are all short-lived gases having a negligible direct GWP component.*
3. *The indirect GWPs are uncertain but could conceivably be comparable in magnitude to the direct GWPs. Only the signs of the indirect effects are estimated here, based on current understanding. The estimates of the indirect effects for the chlorine- and bromine-containing compounds are based on the weight of evidence related to the observed lower stratospheric ozone depletion (WMO, 1992).*
4. *IPCC (1990) contained a typographical error for the indirect effect of methane which led to the inference of an incorrect value for its direct GWP. Also in that report, the indirect effect of NO$_x$ was probably overestimated by a large factor (Johnson et al., 1992).*

Gas	Lifetime (years)	Direct Effect for Time Horizons of 20 years	100 years	500 years	Sign of "Indirect" Effect
CO₂	†	1	1	1	none ††
CH₄	10.5	35	11	4	positive
N₂O	132	260	270	170	uncertain
CFC-11	55	4500	3400	1400	negative
CFC-12	116	7100	7100	4100	negative
CFC-13	400	11000	13000	15000	negative
CFC-14	>500	>3500	>4500	>5300	none †††
HCFC-22	15.8	4200	1600	540	negative
CFC-113	110	4600	4500	2500	negative
CFC-114	220	6100	7000	5800	negative
CFC-115	550	5500	7000	8500	negative
CFC-116	>500	>4800	>6200	>7200	none †††
HCFC-123	1.71	330	90	30	negative
HCFC-124	6.9	1500	440	150	negative
HFC-125	40.5	5200	3400	1200	none †††
HFC-134a	15.6	3100	1200	400	none †††
HCFC-141b	10.8	1800	580	200	negative
HCFC-142b	22.4	4000	1800	620	negative
HFC-143a	64.2	4700	3800	1600	none †††
HFC-152a	1.8	530	150	49	none †††
CCl₄	47	1800	1300	480	negative
CH₃CCl₃	6.1	360	100	34	negative
CF₃Br	77	5600	4900	2300	negative
CHCl₃	0.7	92	25	9	negative
CH₂Cl₂	0.6	54	15	5	negative
CO	months	-	-	-	positive
NMHC	days to months	-	-	-	positive
NO$_x$	days	-	-	-	uncertain

†   The persistence of carbon dioxide has been estimated by explicitly integrating the box-diffusion model of Siegenthaler (1983); an approximate lifetime is 120 years.

†† CO₂ is not involved in chemical reactions affecting the concentrations of the radiatively active species. However, it could affect the relevant chemical reactions through its influences on the atmospheric thermal structure.

††† No currently known or negligible indirect effect.

the total column ozone amounts, the longwave effects are determined both by the amount and its vertical location (Manabe and Strickler, 1964; Ramanathan and Dickinson, 1979; WMO, 1986; Lacis *et al.*, 1990). The loss of ozone in the lower stratosphere induces three distinct radiative effects (Ramanathan *et al.*, 1985; Ramaswamy *et al.*, 1992):

(a) a reduction in the absorption by stratospheric ozone of incoming solar radiation, leading to an increase in the amount reaching the surface-troposphere system;

(b) in the absence of stratospheric temperature changes, a decrease in the emission of longwave radiation from the stratosphere into the surface-troposphere system which is opposite to the solar effect; and

(c) a reduction of the *in-situ* solar heating and a change in the convergence of the longwave radiation within the lower stratosphere. Because the thermal balance of the lower stratosphere is sensitive to radiative perturbations (Fels *et al.*, 1980; Shine, 1987; Kiehl *et al.*, 1988) this third effect results, in the absence of any compensatory dynamical heating, in a decrease of temperatures at these altitudes. This, in turn, leads

to a further reduction of the longwave emission into the troposphere, the magnitude of the reduction being sensitive to the decrease in stratospheric temperatures.

Two sets of radiative transfer modelling studies (WMO, 1992; Ramaswamy *et al.*, 1992), employing slightly different assumptions about the vertical profile of the lower stratospheric ozone loss over the past decade, have been performed and lead to similar conclusions. They suggest an enhancement of the ozone-induced longwave effects (effects b and c) over the solar (effect a) in the middle to high latitudes, corresponding to the larger ozone losses there. Because of the spatial variability in the ozone depletions and the presence of the solar component, the calculated ozone forcing depends on the season and the geographical region. Another factor that has a significant influence on the ozone radiative forcing is the meteorological state of the troposphere, particularly the distribution of clouds (WMO, 1992).

Figure A2.1 (from Ramaswamy *et al.*, 1992) illustrates, for the four seasons and for both hemispheres, the changes in radiative forcing due to ozone, due to CFCs alone, and

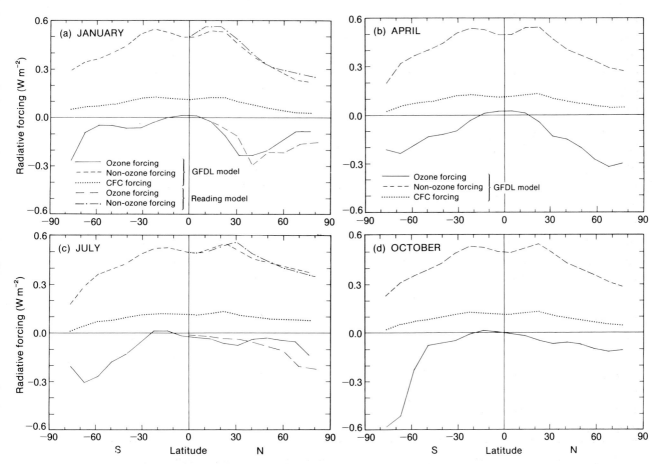

Figure A2.1: Latitudinal and seasonal dependence of the radiative forcing due to (i) the 1979-1990 increases in CFCs, (ii) the 1979-1990 increases in all the non-ozone gases (CO_2, CH_4, N_2O and the CFCs), and (iii) the 1979-1990 observed lower stratospheric losses of ozone (Stolarski *et al.*, 1991; McCormick *et al.*, 1992). Results from two different models are shown: (a) and (c) show University of Reading results (Northern Hemisphere January and July perturbations only), and (a) to (d) show GFDL results. All the results were obtained assuming stratospheric temperature equilibrium in the presence of a fixed dynamical heating (from Ramaswamy *et al.*, 1992).

due to the sum of the non-ozone gases (CO_2, CH_4, N_2O and the CFCs). The illustration indicates that, for the decade of the 1980s, the net ozone radiative forcing in middle to high latitudes is negative, being opposite in sign to the effects due to the non-ozone gases. Poleward of 30 degrees (N and S), the magnitude of the (negative) decadal ozone forcing becomes increasingly comparable to and can even exceed the (positive) CFC forcing over the same time period (WMO, 1992). In higher latitudes, the ozone forcing can counteract a significant fraction of the (positive) total non-ozone gas forcing over the same time period (WMO, 1992). When globally- and annually-averaged (Ramaswamy *et al.*, 1992) and assuming that there is no change in the dynamical heating of the stratosphere, the ozone forcing (-0.08 to -0.09Wm^{-2}), is comparable in magnitude (~80%) but opposite in sign to the decadal CFC greenhouse forcing (0.10 to 0.11Wm^{-2}). The globally-averaged ozone forcing is opposite in sign and is about 18% of the sum of the non-ozone decadal trace gas forcing (0.45 to 0.47Wm^{-2}).

It is emphasized that the ozone forcing is extremely sensitive to the altitude of the losses (Ramanathan *et al.*, 1985; Lacis *et al.*, 1990). There still is some uncertainty regarding the exact profile and the magnitude of the loss in the immediate vicinity of the tropopause. The SAGE profiles are available globally only from ~17km and above, and suggest an increasing percentage of loss with decreasing altitude in the lower stratosphere. As an illustration of this sensitivity, let us suppose that the losses observed by TOMS are uniformly distributed through the entire stratospheric column (see WMO, 1986). For this to be the case, the principal ozone depletions would have to occur at altitudes higher than observed over the past decade, in which case a much smaller global ozone forcing (-0.01 to -0.04Wm^{-2}) would result. Thus, inferences about ozone forcing depend crucially on both the total column change as well as the change in the vertical profile.

The radiative forcing due to ozone is unique in two respects when compared to that due to the non-ozone gases. First, although there is an approximate global mean offset of the direct CFC forcing by the ozone losses occurring during the 1979-1990 period, this arises because of a significant negative forcing by ozone occurring only in the middle to high latitudes, in particular the radiative forcing due to increasing CFCs and decreasing ozone ranges from a net positive one at low latitudes to a net negative one at higher latitudes for the period considered (WMO, 1992). Because of the spatial dependence of the ozone losses and consequently the ozone forcing, the globally averaged results represent a considerable simplification and must be treated with caution.

Second, for ozone, unlike the other radiatively active species, both solar and the longwave interactions become significant. The negative surface-troposphere forcing at the mid-to-high latitudes consists of a solar-induced warming tendency at the surface, combined with a longwave-induced cooling tendency of the troposphere (Ramanathan and Dickinson, 1979; WMO, 1992).

A2.4.2 The Greenhouse Effect of Ozone Losses

The weight of evidence suggests that the observed stratospheric ozone losses are due to heterogeneous chemical reactions involving chlorine- and bromine-containing chemicals (halocarbons), particularly the anthropogenic emissions of CFCs (WMO, 1992). The computed ozone radiative forcings indicate that the indirect chemical effects due to the halocarbons have substantially reduced the radiative contributions of the CFCs to the greenhouse forcing over the past decade (WMO, 1992). Thus, the net greenhouse impact attributed to the CFCs taken together, including the indirect as well as direct contributions, may be significantly reduced because of the halocarbon-induced destruction of ozone.

Three-dimensional General Circulation Model (GCM) simulations of the impacts on the Earth's climate due to the ozone losses have not been performed as yet, so estimates of the effects on the surface temperature are not available. In an investigation of the climatic effects due to the Northern Hemisphere mid-latitude ozone changes around the tropopause during the decade of the 1970s, Lacis *et al.* (1990) estimated a cooling of the surface at those latitudes. One-dimensional globally- and annually-averaged radiative-convective models indicate that, while a loss of ozone in the lower stratosphere leads to a surface cooling, ozone losses in the middle and upper stratosphere, as predicted from homogeneous gas-phase chemistry models, yield a warming of the surface (Ramanathan *et al.*, 1985).

A2.4.3 Ozone Depletion and Stratospheric Temperature Changes

The indirect effect of CFCs on the climate system due to depletion of ozone in the lower stratosphere is critically sensitive to the actual temperature change and its distribution in the lower stratosphere. Because atmospheric circulation can change in response to radiative perturbations, the dynamical contribution to the heating could also change, thereby contributing to the actual temperature change in the stratosphere (Dickinson, 1974).

The observed global ozone depletion has not yet been simulated in a GCM, but simulations for the following scenarios of ozone changes have been performed: (a) a uniform decrease of O_3 throughout the stratospheric column (Fels *et al.*, 1980; Kiehl and Boville, 1988), (b) a homogeneous gas-phase chemical model prediction of ozone depletion (Kiehl and Boville, 1988), which is different from the observed losses, and (c) observed springtime depletion in the Antarctic region (Kiehl *et al.*,

1988). The resulting stratospheric temperature changes in these studies indicate that, unless the column depletions are large (>50%), the GCM response is similar to the response indicated by the fixed dynamical heating model but there are some latitude-dependent differences. The GCM studies of Rind *et al.* (1990, 1991) suggest that dynamically forced temperature changes can result from subtle interactions between changes in the atmospheric structure, upper tropospheric latent heat release, and the forcing and transmission of planetary waves and gravity waves.

Turning to the observed temperature trend (see Section C3.3), the long-term temperature trends in the lower stratosphere suggest a cooling. While this would be in accord with the ozone-induced radiative cooling tendency of the lower stratosphere (Lacis *et al.*, 1990; WMO, 1992; Miller *et al.*, 1992) considerable difficulties remain in attributing a part or all of the observed stratospheric temperature trends to the ozone losses. Other physical factors, such as possible changes in the physical state of the troposphere, volcanic aerosols, change in stratospheric circulation, etc., which are not accounted for in the determination of the radiative forcing, could also be contributing to the trends. Thus, significant questions remain with regard to the influence of ozone losses on stratospheric temperatures.

A2.5 Radiative Forcing due to Gases in the Troposphere

A2.5.1 Introduction

Indirect greenhouse gas effects, induced by changes in the chemistry of the troposphere, are likely to be significant. Gases which are key compounds for the indirect effects (and for the oxidation processes in the troposphere) are O_3 and OH.

Ozone is a greenhouse gas by itself. It is formed *in situ* in the atmosphere by photochemical processes. Tropospheric O_3 concentrations are influenced by the distribution of CH_4, CO, NO_x and NMHCs, leading to indirect greenhouse contributions for these gases. OH is of importance for the greenhouse effect because it controls the loss of a large number of greenhouse gases in the troposphere (CH_4, HFCs, HCFCs, etc.), thereby determining their chemical lifetimes. Furthermore, O_3 and OH will be affected by the release of these gases, leading to important feedback effects on their lifetimes. Additional indirect greenhouse effects arise from CH_4, CO and NMHCs since CO_2 is the end product of their chemical oxidation in the troposphere.

Since the tropospheric chemical processes determining the indirect greenhouse effects are highly complex and not fully understood, the uncertainties connected with estimates of the indirect effects are larger than the

uncertainties of those connected to estimates of the direct effects. Added to these uncertainties are the limitations of the current models in formulating the distribution and spatial variability of NO_x in the troposphere, or the impact of clouds on gas phase chemistry (see Chapter 5, WMO, 1992).

This sub-section examines the key processes for O_3 and OH formation, the role of gaseous emissions of CH_4, NO_x, CO and NMHC in changing the O_3 and OH distributions (and thereby leading to indirect effects), and uncertainties in estimating these impacts. The impact of increased UV fluxes in the troposphere due to reduced ozone columns, and enhanced water vapour content resulting from enhanced temperatures are not included. Both effects are likely to enhance the OH levels in the troposphere.

A2.5.2 Chemical processes and changes in O_3 and OH

Although tropospheric O_3 only makes up about 10% of all ozone in the atmosphere, its presence is central to the problem of the oxidizing efficiency of the troposphere. O_3 photolysis is the primary source of OH radicals as well as being an oxidizing species itself. Through the formation of OH it determines the cleansing efficiency of the troposphere. Ultimately, therefore, ozone is one of the most important constituents in determining the chemical composition of the troposphere.

The production of ozone depends on the concentrations of NO_x (WMO, 1992, Chapter 5). As the latter compounds are short-lived, and their concentrations vary strongly in the troposphere, ozone production is believed to vary significantly throughout the troposphere. There is also an *in situ* chemical loss of ozone in the troposphere in areas where NO_x concentrations are particularly low (less than approximately 20 pptv). This occurs in the middle troposphere and in remote oceanic areas where there are no NO_x sources (Liu *et al.*, 1987). Ozonesonde measurements reported in WMO (1992) indicate that ozone has increased by 1-1.5% per year in the free troposphere over Europe during the last 20 to 25 years (Staehelin and Schmid, 1991). A similar trend in tropospheric ozone has previously been reported for stations influenced by regional air pollution (Logan, 1985; Bojkov, 1987; Penkett, 1988).

Methane is oxidized primarily in the troposphere (>90%) through reaction with the hydroxyl radical. Since this reaction also provides a substantial fraction of the OH loss in the troposphere, there is a strong interaction between OH and CH_4. This causes OH to decrease when CH_4 increases, leading to a further increase in CH_4 - a positive feedback (Chameides *et al.*, 1977; Sze, 1977; Isaksen, 1988).

Because of the central role O_3 and OH play in tropospheric chemistry, the chemistry of CO, CH_4, NMHC and NO_x is strongly intertwined, making the interpretation of emission changes rather complex. In general, increases in the emissions of hydrocarbons and of CO may lead to increases in the global average O_3, but to reductions in OH levels (Isaksen and Hov, 1987). The result will be a slower atmospheric loss of methane and other species controlled by OH (e.g., HFCs, HCFCs). The impact of NO_x changes is different. Increased emissions of NO_x lead to increases in both globally averaged O_3 and OH (Isaksen and Hov, 1987). The increased OH levels reduce the lifetime of methane. The important consequence of this is that NO_x emissions have opposing effects on the two greenhouse gases O_3 and CH_4.

Significant changes in the global distribution of OH may have occurred over the last two centuries as the trace gas composition of the troposphere has changed dramatically. Key compounds like CH_4 and CO have increased in concentrations. This is expected to have led to reduced OH levels. On the other hand, increases in the concentrations of NO_x that are believed to have occurred (although it has not been possible to measure any changes) will have tended to reduce OH levels. The net effect is difficult to estimate, and there are no direct measurements of OH which can give reliable information on the global distribution or changes over time. Several indirect methods have, however, been used to derive a global distribution of OH (WMO, 1992). A recent estimate of the OH trend is the analysis of Prinn *et al.* (1992). They deduce a trend in global OH from the ALE/GAGE CH_3CCl_3 record of $+1.0\pm0.8\%$ per year over the past decade. Their result was based on a simple tropospheric box model.

A2.5.3 Sensitivity of Radiative Forcing to Changes in Tropospheric Ozone

The radiative forcing due to increases in tropospheric ozone has been investigated in earlier reports (WMO, 1986). Even though tropospheric ozone amounts are less than in the stratosphere, their effective longwave optical pathlength is comparable to that in the stratosphere (Ramanathan and Dickinson, 1979), thus rendering them radiatively important. In particular, the upper tropospheric concentrations are most significant (Lacis *et al.*, 1990). A 10% uniform increase with height in the concentrations of tropospheric ozone from current levels at 40°N (January conditions) yields a positive radiative forcing of about $0.1 Wm^{-2}$ (WMO, 1992). Although the radiative effect of increases in tropospheric ozone could be extremely important in the greenhouse forcing of climate, there is, at present, insufficient evidence that such increases are actually taking place globally, especially in the radiatively significant upper tropospheric regions. In the absence of data from which meaningful global trends can be derived, it is not possible at this stage to quantify the current contribution of tropospheric ozone to the global greenhouse radiative forcing.

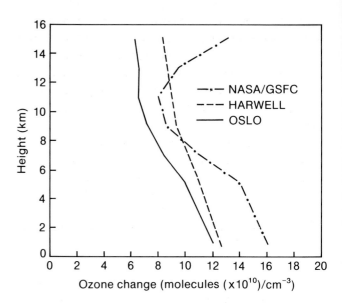

Figure A2.2: Calculated global average ozone increases due to a doubling (compared to current levels) of surface emissions of methane (2-D model calculations with the Harwell and the Oslo models refered to in WMO, 1992) and a doubling of the surface concentrations of methane (1-D model calculations with the NASA/GSFC model). In the latter case, ozone production is most likely underestimated compared to the other two cases because it implies smaller emission increases (see discussion in the text).

A2.5.4 Indirect Effects due to CH_4 Emissions
A2.5.4.1 Changes in Lifetimes due to Changes in OH
Enhanced surface emissions of CH_4 cause increased ozone levels which show moderate variations with latitude and season (WMO, 1992). The increase is most pronounced in the lower troposphere and decreases with height. Figure A2.2 shows calculated global and seasonal average ozone increases (in molecules per cm^3) with height for three different tropospheric models resulting from a doubling of surface CH_4 (fluxes and concentrations).

The positive feedback on methane through the impact on the OH distribution is found to be substantial. Furthermore, the feedback is non-linear: it increases with increasing CH_4 emission in the sense that the relative increase in steady-state concentration for a given increase in emissions increases faster than the relative increase in emissions. For example, a 10% increase in emission leads to a 13 to 14% increase in steady-state concentration, while a doubling of emissions leads to a 150% increase in concentration. In a similar way the feedback will affect the concentrations when emissions are reduced, but it will become less significant at lower concentrations.

A2.5.4.2 Radiative Forcing Changes from CH_4-Induced Changes in Ozone
Methane has an indirect effect on the radiation balance through its influence on ozone. Ozone changes in the troposphere due to increased CH_4 surface fluxes (Figure

A2.2) have been used to calculate the change in radiative forcing at different latitudes and seasons using a radiative-convective model (W.-C. Wang, personal communication). The relative effect of ozone on the long wavelength radiation (compared to the effect of CH_4 alone) varies between 10 and 22%. The globally-averaged effect is around 14%.

The CH_4 effect in these calculations includes the direct effect and the indirect effect resulting from reduced OH levels (the positive feedback effect). Approximately 0.7 of the methane effect is a direct effect, and 0.3 is due to the OH feedback. The calculated indirect global effect through ozone increases is therefore approximately 20% of the direct methane effect.

A2.5.4.3 Effects of CH_4 on Stratospheric Water Vapour

Methane has an indirect effect through its oxidation in the stratosphere to water vapour. In IPCC (1990) it was assumed that this would enhance the methane forcing by 30% over its direct value in the absence of N_2O absorption-band overlap (see Table 2.2 in IPCC, 1990). More recent modelling studies have shown that this enhancement is highly uncertain. In WMO (1992) the range is 22 to 38%, while Lelieveld and Crutzen (1992) give a value of 5%. These differences may partly reflect the different types of numerical experiment performed to calculate the effect. They may also reflect the fact that changes depend critically on the vertical profile of the water vapour change (A.A. Lacis, personal communication).

A2.5.4.4 Oxidation to CO_2

Oxidation of CH_4 leads to formation of CO_2 and thus contributes indirectly to greenhouse warming. The contribution will depend on the time horizon used. It should be noted that only oxidation of fossil fuel-related CH_4 leads to a CO_2 increase; most CH_4 emitted into the atmosphere is short-term recycled biogenic CO_2.

A2.5.4.5 Indirect GWP for CH_4.

While the results given here demonstrate the importance of a number of processes in amplifying the direct radiative forcing effect of increasing methane concentrations, they cannot be applied to directly scale up the direct GWP for methane. The experiments performed give only the steady-state changes due to a sustained emission change, whereas the GWP definition used here considers the integrated time-dependent response to a pulse emission.

A2.5.5 NO_x Emissions

The calculated global ozone changes from increases in NO_x surface emissions show large seasonal, latitudinal and height variations (WMO, 1992). The impact on ozone drops off rapidly with height in the troposphere. This is significant as the impact on surface temperatures from ozone changes is likely to be height dependent in the troposphere with the largest effect resulting from changes in the upper troposphere (Wang and Sze, 1980; Lacis *et al.*, 1990). Furthermore, the calculations show that the results are highly model sensitive, leading to large differences (more than a factor of 2) in ozone impact between the 2-D models used.

Calculation of the impact on ozone from NO_x emissions from aircraft indicate that this source of NO_x may be more than an order of magnitude more efficient in enhancing O_3 levels than surface emissions of NO_x. The enhancement occurs also at higher altitudes (middle and upper troposphere) where O_3 changes have larger effects on surface temperatures (Wang and Sze, 1980; Lacis *et al.*, 1990).

Increased NO_x levels are expected to increase global amounts of OH, and thus to lead to a reduction in CH_4. NO_x increases therefore have an opposite effect on the abundance of the two greenhouse gases, O_3 and CH_4. Calculations also indicate that the radiative forcings caused by NO_x-induced changes in O_3 and CH_4 could be of the same magnitude. Taking all this into account, estimates of the radiative forcing of NO_x changes are extremely uncertain and cannot be reliably made at present.

There is a consistent picture emerging from the calculations of OH sensitivity to enhanced fluxes of source gases showing that increased NO_x emissions lead to increased OH and thereby to increased oxidation in the troposphere, while increases in the emissions of the other source gases lead to reduced OH values.

A2.5.6 CO and NMHC Emissions

Ground-based emissions of CO and NMHC lead to O_3 production, but these are found to be less efficient ozone producers on a global scale than CH_4. OH is also increased. The increases show pronounced global and seasonal variations making estimates of indirect GWP highly uncertain.

A2.5.7 Summary

A summary of the indirect contributions to radiative forcing from CH_4, NO_x, CO and NMHC is given in Table A2.2. The Table gives the sign of the contribution from the individual processes and the net effect, but no absolute values for the indirect effects are presented.

For CH_4, all the indirect contributions are believed to be positive and therefore they add to the direct GWP. The sum of their contributions is likely to be significant, possibly similar in magnitude to the direct effect. Indirect effects for methane will thus add substantially to the direct values given in Table A2.1. CO and NMHC will also make positive indirect contributions, although they are believed to be less significant than the contribution from CH_4 and

Table A2.2: *Summary of the impact of emissions of CH$_4$, NO$_x$, CO and NMHC on OH, on O$_3$ and those other greenhouse gases affected by changes in OH*

	Effect on concentrations of		
Increased emission of:	OH	O$_3$	CH$_4$, HCFCs, HCFs
CH$_4$	-	+	+
NO$_x$	+	+	-
CO	-	+	+
NMHC	-	+	+

more difficult to assess due to temporal and spatial variations in concentrations. For NO$_x$, uncertainties are large and it is not possible to give even the sign of the indirect effect. It should be noted, however, that NO$_x$ emitted from aircraft in the upper troposphere could have a large impact on the chemistry and, thereby, on the radiative forcing on the surface-troposphere-system.

Although indirect GWP values were given in IPCC (1990), we are now aware of additional complications affecting such calculations and are less sure of the results. As a consequence, no indirect GWP values are given in this report. For methane, however, the indirect contributions to the GWP are likely to be significant, possibly as large as the direct effect.

A2.6 Radiative Forcing due to Aerosol Particles

A2.6.1 Tropospheric Particles

A2.6.1.1 Background

Aerosol particles influence the Earth's radiative balance directly by scattering and absorption of shortwave (solar) radiation. An increase in concentrations of aerosol particles will enhance this effect. Aerosol particles also absorb and emit longwave (infrared) radiation, but this effect is usually small because (a) the opacity of aerosols decreases at longer wavelengths and (b) the aerosols are most concentrated in the lower troposphere where the atmospheric temperature, which governs emission, is practically the same as the surface temperature (Coakley *et al.*, 1983; Grassl, 1988).

Aerosol particles also serve as sites on which cloud droplets form (cloud condensation nuclei, CCN). Increased concentrations of aerosol particles have the potential, therefore, to alter the microphysical, optical and radiative properties of clouds, changing their reflective properties and possibly their persistence.

By providing additional surfaces for condensation and heterogeneous chemistry, aerosol particles may also influence the chemical balance of gaseous species in the atmosphere. This may be especially important for the balance of O$_3$ in the lower stratosphere.

In the unperturbed atmosphere, the principal aerosol constituents contributing to light scattering are sulphate from biogenic gaseous sulphur compounds and organic carbon from partial atmospheric oxidation of gaseous biogenic organic compounds. Seasalt and windblown soil dust contribute substantially at some locations but their effect on the global climate is generally unimportant because the particles are large and usually short-lived and thus transported only short distances. Other aerosol substances may also be locally and regionally important, especially those that are sporadic such as from volcanoes, wildfires, and windblown dust from deserts.

Submicrometre (diameter $<10^{-6}$m) anthropogenic aerosols are produced in the atmosphere by chemical reactions of primarily sulphur-, but also nitrogen- and carbon-containing gases, predominantly sulphur dioxide. These particles are efficient light scatterers, so changes in them may alter the energy balance of the atmosphere. The overall effect also depends on changes in the amount of energy absorbed. Light absorption is dominated by particles containing elemental carbon produced by incomplete combustion of carbonaceous fuel. The light scattering effect is dominant at most latitudes, but absorption might dominate at high latitude, especially over highly reflective snow- or ice-covered surfaces (Blanchet, 1989). Over surfaces with low albedo (<0.1) characteristic of most of the surface of the Earth, anthropogenic aerosols cool rather than warm the Earth (Coakley *et al.*, 1983); thus in global terms the scattering by sulphate should dominate over absorption by elemental carbon.

A2.6.1.2 New findings

The most important new information that has become available since IPCC (1990) refers to the backscattering by sulphate aerosols. Based on simulations of the global distribution of sulphate aerosols (Langner and Rodhe, 1991), Charlson *et al.* (1990, 1991 and 1992) used previously acquired information on scattering and backscattering coefficients per unit mass of sulphate to estimate the impact of anthropogenic sulphur emissions on the shortwave radiation balance in cloud-free regions. The

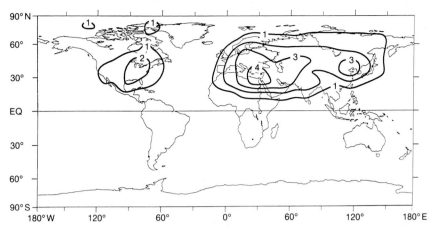

Figure A2.3: Calculated increase of reflected flux to space due to tropospheric sulphate aerosols derived from anthropogenic sources. Unit: Wm^{-2} (from Charlson *et al.*, 1991).

geographical distribution of the calculated increase of reflected solar flux to space is shown in Figure A2.3. The authors concluded that the effect for current emission levels, averaged over the Northern Hemisphere, corresponds to a negative forcing at the Earth's surface of about 1Wm^{-2}, with about a factor of two uncertainty. This is comparable (but of opposite sign) to the forcing due to anthropogenic CO$_2$ (+1.5Wm^{-2}) or to the direct forcing of the other greenhouse gases (+1Wm^{-2}). In addition to the direct effect on climate of sulphate aerosols, there is an indirect effect - via changes in CCN and cloud albedo - which tends to act in the same direction (i.e., towards a cooling) with a magnitude that has not yet been reliably quantified (Charlson *et al.*,1990 and 1992; Kaufman *et al.*, 1991).

A very important implication of this estimate is that the net anthropogenic radiative forcing over parts of the Northern Hemisphere during the past century is likely to have been substantially smaller than was previously believed. A quantitative comparison between the positive forcing of the greenhouse gases and the negative forcing due to sulphate is complicated by the fact that the latter is much less uniformly distributed geographically than the former, cf. Figure A2.3.

A2.6.1.3 Discussion

Future changes in the forcing due to sulphate aerosols and greenhouse gases will depend on how the corresponding emissions vary. Because of the short atmospheric lifetimes of sulphate and its precursors, atmospheric concentrations will adjust within weeks to changes in emissions. This is a very different situation from that for most greenhouse gases which have effective lifetimes of decades to centuries. For example, the concentration of CO$_2$ will continue to rise for more than a century even if emissions are kept constant at today's level. This difference is illustrated in Figure A2.4 (from Charlson *et al.*, 1991), which shows schematically

how the climate forcings due to CO$_2$ and aerosol sulphate could change if the global fossil fuel consumption levelled off and eventually were reduced. More detailed calculations have been given by Wigley (1991). Because of the rapid growth in emissions during the past decades both the enhanced greenhouse forcing due to CO$_2$ and the opposite forcing due to aerosol sulphate have grown

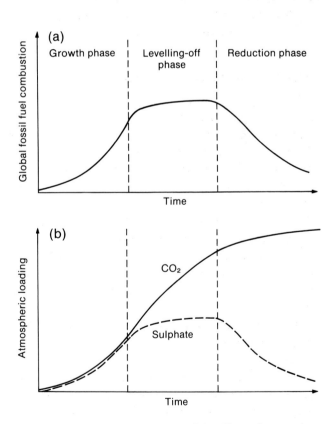

Figure A2.4: Schematic illustration of the effects of a scenario of future global fossil fuel combustion (a) on the atmospheric loadings of CO$_2$ and sulphate (b). The differences in response arise from their different atmospheric residence times. CO$_2$ produces a positive radiative forcing (heating) and sulphate a negative forcing (cooling) (from Charlson *et al.*, 1991).

accordingly. During a levelling off phase, the greenhouse forcing due to CO_2 will continue to grow whereas the aerosol forcing will remain constant. During a decay phase, the greenhouse forcing will start to level off and the aerosol forcing decline. This simple example illustrates that the relative importance of these two major anthropogenic forcing agents in the future will depend critically on the character of changes in fossil fuel use (large-scale desulphurization measures would also have to be considered).

Because of the very different character of the forcing due to aerosol sulphate as compared to that of the greenhouse gases, no attempt is made here to define a negative Global Warming Potential (GWP) for anthropogenic sulphur emissions. In addition to the well-known negative deleterious environmental effects of sulphates - including acidification - there are several reasons why increases in sulphur emissions cannot simply be considered as trade-offs against reductions in greenhouse gas emissions. Among these are: the very different horizontal and vertical distributions of radiative forcing due to sulphates compared with greenhouse gases, the great uncertainty about their effects on clouds and the subsequent effects on the climate system, and the fact that the distribution of sulphates globally is largely inferred from models rather than being directly measured. Although in a global sense the negative forcing due to aerosols may offset a substantial part of the positive greenhouse forcing, the differences in spatial distribution of greenhouse radiative forcing and aerosol effects mean that increases in sulphates can never be expected to compensate for the climatic effects of greenhouse gas increases.

It is clear that a better quantitative description of the climate influence of anthropogenic aerosols is necessary in models of past, present and future climate. Modelling studies need to take into account sulphate aerosol concentrations and their radiative influence as a function of location and time as governed by emissions of sulphur gases. To do this, models should accurately represent the direct light scattering effect and also the influences of these aerosols on cloud optical, radiative, and persistence properties. At present, the influences on clouds in particular cannot be estimated with confidence. Further, there is only a meagre data base of observations with which to validate the models of atmospheric chemistry, transport and removal processes that are required to relate aerosol concentrations to precursor emissions. There is, similarly, little observational information on the relationships between aerosol microphysical properties and cloud microphysical properties, between aerosol and cloud microphysical properties and their radiative properties, and between cloud microphysical properties and cloud persistence.

A2.6.2 Stratospheric Sulphate Particles

Observations over the past decade (lidar, satellite, balloon, sunphotometer) indicate that the stratospheric aerosol concentration throughout most of the 1980s remained higher than that measured in 1979 (a relatively quiescent period). This is probably largely attributable to the major El Chichon volcanic eruption in 1982 together with the effects of a few other minor eruptions (McCormick and Trepte, 1987). Anthropogenic sources may have provided an additional contribution (Hofmann, 1990). With the recent major eruption of the Mt. Pinatubo volcano, there is now a fresh accumulation of particulates in the stratosphere (optical depth estimated to be between 0.1 and 0.3 one month after the eruption; M. P. McCormick, personal communication). For more details, the reader is referred to the special issue of *Geophysical Research Letters* (Vol. 19, 149-218, 1992).

The radiative effects due to these particles may be already manifest in an observed warming of the lower stratosphere (see Section C4.2.4.2; also Labitzke and McCormick, 1992). Radiative forcing calculations indicate that these aerosols can also be expected to exert a significant negative but transient radiative forcing ($\sim 0.5 Wm^{-2}$ or more in magnitude) on the surface-troposphere system over the next few years (Hansen *et al.*, 1992). This is opposite in sign to the greenhouse gas-induced forcing. General Circulation Model simulations (Hansen *et al.*, 1992) suggest that such an aerosol-induced forcing could yield a temporary cooling tendency at the surface and dominate the global surface temperature record in the next year or more (see Section C4.2.4.2).

A2.7 Forcing Due to Solar Irradiance Changes

For a recent review of changes in solar irradiance, see Lean (1991). A 1% change in total irradiance is equivalent to a radiative forcing of $2.4 Wm^{-2}$ at the top of the troposphere, comparable to the total enhanced greenhouse forcing to date. Considerably smaller changes than this could, if sustained for a number of years, noticeably affect global climate and either enhance or offset the effects of increasing greenhouse gas concentrations. It is necessary, therefore, to monitor future and, if possible, reconstruct past irradiance changes with an accuracy of substantially better than ±1%.

To obtain better accuracy, it is necessary to place instruments high in or above the atmosphere. Continuous observations require satellite instrumentation and the available record spans only the period from 1978 to the present. These data show a strong link between solar magnetic activity (sunspots, faculae and the background "active" network radiation) and total irradiance on time-scales of days to years. During the sunspot minimum of 1986, total irradiance was about 0.1% less than during the previous (1980) maximum, with the reduction in output from bright features outweighing the decreased blocking effect of sunspots (Foukal and Lean, 1988, 1990; Lean, 1989). Changes since 1986 have continued to parallel the 11-year sunspot cycle (Willson and Hudson, 1991).

The small amplitude of the observed solar-cycle-related irradiance changes does not preclude the existence of additional lower-frequency effects operating on the 10 to 100 year time-scale (Foukal and Lean, 1990; Lean, 1991). To investigate this possibility further, there have been attempts to extend the observational record back before 1980 by using rocket- and balloon-based measurements (Fröhlich, 1987). These show an apparent change in irradiance between the late 1960s and the late 1970s of around 0.4%. There is considerable doubt, however, about the representativeness of these values (measured over time-scales of a day) and about the absolute accuracy of the instruments used to obtain them (Lean, 1991). Although Reid (1991) argues against these problems, it is clearly difficult to identify a long-term trend using extremely noisy daily data from instruments of uncertain accuracy.

Apart from these data, there are no useful direct irradiance measurements prior to 1978, so various authors have tried to deduce irradiance forcing indirectly. For example, Reid (1991) has suggested that low-frequency irradiance changes run parallel to the envelope of sunspot activity, which shows quasi-cyclic behaviour with a roughly 80-year period, and Friis-Christensen and Lassen (1991) have hypothesized that low-frequency irradiance changes are related to changes in the length of the solar cycle. In both cases, there is a strong visual correspondence between the solar irradiance proxy and global-mean temperature changes over the past 100 years - see Section C4.2.1. These results are intriguing, but they have yet to be fully evaluated in terms of the implied changes in solar forcing compared to greenhouse forcing (Kelly and Wigley, 1990).

An entirely different approach has been used in a study by Baliunas and Jastrow (1990) - see also Radick *et al.* (1990). They have examined the magnetic activity of solar-type stars to try to throw some light on possible changes in irradiance associated with events like the Maunder Minimum of sunspot activity (1645-1715). The precisely-dated record of atmospheric radiocarbon measurements shows that similar periods of prolonged sunspot minima have occurred on many occasions during the past 8000 years (randomly spaced, but every 500 years on average). It has been suggested that they correspond to periods of lowered irradiance and global cooling (Eddy, 1976; Wigley and Kelly, 1990). Baliunas and Jastrow (1990) find that solar-type stars exhibit two modes of activity, a cyclic mode similar to the Sun's present condition, and a less variable mode (with lower magnetic activity) akin to

conditions thought to prevail during the Maunder Minimum. They conclude that the Sun's irradiance during the Maunder Minimum (and other similar periods) was "several tenths of a per cent" less than current levels.

A more sophisticated interpretation of these stellar data has been carried out by Lean *et al.* (1992) using knowledge of the mechanisms of irradiance variations gleaned from extant solar data. They consider two possible effects: changes in irradiance associated directly with changes in magnetic activity, and changes associated with a reduced basal emission during prolonged periods of reduced activity. The estimated irradiance reduction during a Maunder Minimum period is $0.25\pm0.1\%$.

It should be noted that these astronomical results do not yet prove that the Sun's irradiance was reduced during periods like the Maunder Minimum. Since no star has been observed to change mode, it is not yet known whether the observed stellar differences reflect different types of star or different modes of variation for individual stars. Nevertheless, the magnitude of the potential changes estimated by Lean *et al.* (1992) and Baliunas and Jastrow (1990) compares favourably with the empirical estimate (based on palaeoclimatic data) for a Maunder Minimum irradiance reduction of 0.22-0.55% given by Wigley and Kelly (1990). All three estimates are substantially below that of Reid's (1991) estimate of around 1%. Wigley and Kelly (1990) note that, were a similar event to begin now or in the near future, then it would partially offset the anticipated increase in forcing due to increasing greenhouse gas concentrations but only by a small amount.

References

Baliunas, S. and R. Jastrow, 1990: Evidence for long-term brightness changes of solar-type stars. *Nature,* **348,** 520-523.

Blanchet, J.-P, 1989: Toward estimation of climate effects due to Arctic aerosols. *Atmos. Env.,* **23,** 2609-2625.

Bojkov, R.D., 1987: Ozone changes at the surface and in the free troposphere. In: *Tropospheric Ozone.* Proceedings of the NATO Workshop. I.S.A. Isaksen (Ed.), D. Reidel, Boston, pp83-96.

Chameides, W.L., S.C. Liu and R.J. Cicerone, 1977: Possible variations in atmospheric methane. *J. Geophys. Res.,* **82,** 1795-1798.

Charlson, R.J., J. Langner and H. Rodhe, 1990: Sulfate aerosols and climate. *Nature,* **348,** 22-26.

Charlson, R.J , J. Langner, H. Rodhe, C.B. Leovy and S.G. Warren, 1991: Perturbation of the Northern Hemisphere radiative balance by backscattering from anthropogenic sulfate aerosols. *Tellus,* **43A-B** (4), 152-163.

Charlson, R.J, S.E. Schwartz, J.M. Hales, R.D. Cess, J.A. Coakley Jr., J.E. Hansen and D.J. Hofmann, 1992: Climate forcing by anthropogenic aerosols. *Science,* **255,** 423-430.

Coakley, J.A. Jr., R.D. Cess and F.B. Yurevich, 1983: The effect of tropospheric aerosols on the Earth's radiation budget: a parametrization for climate models. *J. Atmos. Sci.*, **40**, 116-138.

Dickinson, R. E., 1974: Climate effects of stratospheric chemistry. *Can. J. Chem.*, **52**, 1616-1624.

Eddy, J.A., 1976: The Maunder Minimum. *Science*, **192**, 1189-1202.

Fels, S.B., J.D. Mahlman, M.D. Schwartzkopf and R.W. Sinclair, 1980: Stratospheric sensitivity to perturbations in ozone and carbon dioxide: Radiative and dynamical response. *J. Atmos. Sci.*, **37**, 2266-2297.

Fisher, D.A., C.H. Hales, W.-C. Wang, M.K.W. Ko and N.D. Sze, 1990: Model calculations of the relative effects of CFCs and their replacements on global warming. *Nature*, **344**, 513-516.

Foukal, P. and J. Lean, 1988: Magnetic modulation of solar luminosity by photospheric activity. *Astrophysical Journal*, **328**, 347-357.

Foukal, P. and J. Lean, 1990: An empirical model of total solar irradiance variation between 1874 and 1988. *Science*, **247**, 556-558.

Friis-Christensen, E. and K. Lassen, 1991: Length of the solar cycle: an indicator of solar activity closely associated with climate. *Science*, **254**, 698-700.

Fröhlich, C., 1987: Variability of the solar "constant" on time scales of minutes to years. *J. Geophys. Res.*, **92**, 796-800.

Grassl, H., 1988: What are the radiative and climatic consequences of the changing concentration of atmospheric aerosol particles? In: *The Changing Atmosphere*. F.S. Rowland and I.S.A. Isaksen (Eds.), Wiley and Sons, Chichester, pp187-199.

Hansen, J., A. Lacis, R. Ruedy and M. Sato, 1992: Potential climate impact of Mount Pinatubo eruption. *Geophys. Res. Lett.*, **19**, 215-218.

Hofmann, D.J., 1990: Increase in the stratospheric background sulfuric acid aerosol mass in the past 10 years, *Science*, **248**, 996-1000.

IPCC, 1990: *Climate Change: The Scientific Assessment*. J.T. Houghton, G.J. Jenkins and J.J. Ephraums (Eds.), Cambridge University Press, Cambridge, UK, pp365.

Isaksen, I.S.A. and Ø.Hov, 1987: Calculation of trends in the tropospheric concentration of O_3, OH, CO, CH_4 and NO_x. *Tellus*, Ser. B., **39**, 271-285.

Isaksen, I.S.A., 1988: Is the oxidizing capacity of the atmosphere changing? In: *The Changing Atmosphere*. F.S. Rowland and I.S.A. Isaksen (Eds.), Wiley-Interscience, pp141-157.

Johnson, C., J. Henshaw and G. McInnes, 1992: Impact of aircraft and surface emissions of nitrogen oxides on tropospheric ozone and global warming. *Nature*, **355**, 69-71.

Kaufman, Y.J., R.S. Fraser and R.L. Mahoney, 1991: Fossil fuel and biomass burning effect on climate – heating or cooling? *J. Clim.*, **4**, 578-588.

Kelly, P.M. and T.M.L. Wigley, 1990. The influence of solar forcing trends on global mean temperature since 1861. *Nature*, **347**, 460-462.

Kiehl, J.T., B.A. Boville and B.P. Briegleb, 1988: Response of a general circulation model to a prescribed Antarctic ozone hole. *Nature*, **332**, 501-504.

Kiehl, J.T. and B.A. Boville, 1988: The radiative-dynamical response of a stratospheric-tropospheric general circulation model to changes in ozone. *J.Atmos.Sci.*, **45**, 1798-1817.

Labitzke, K. and M.P. McCormick, 1992: Stratospheric temperature increases due to Pinatubo aerosols. *Geophys. Res. Lett.*, **19**, 207-210.

Lacis, A.A., D.J. Wuebbles and J.A. Logan, 1990: Radiative Forcing of Climate by Changes in the Vertical Distribution of Ozone. *J. Geophys. Res.*, **95**, D7, 9971-9981.

Langner, J. and H. Rodhe, 1991: A global three-dimensional model of the tropospheric sulfur cycle. *J. Atmos. Chem.* **13**, 225-263.

Lean, J., 1989: Contribution of ultraviolet irradiance variations to changes in the Sun's total irradiance. *Science*, **244**, 197-200.

Lean, J., 1991: Variations in the Sun's radiative output. *Rev. Geophys.*, **29**, 505-535.

Lean, J., W. Livingston, A. Skumanich and O. White, 1992: Estimating the Sun's radiative output during the Maunder Minimum. (Unpublished manuscript).

Lelieveld, J. and P.J. Crutzen, 1992: Indirect effects of methane in climate warming. *Nature*, **355**, 339-342.

Liu, S.C., M. Trainer, F.C. Fehsenfeld, D.D. Parrish, E.J. Williams, D.W. Fahey, G. Hubler and P.C. Murphy, 1987: Ozone production in the rural troposphere and the implications for regional and global ozone distribution. *J. Geophys. Res.*, **92** (4), 191-207.

Logan, J.A., 1985: Tropospheric ozone: seasonal behavior, trends, and anthropogenic influence. *J. Geophys. Res.*, **90**, 10463-10482.

McCormick, M.P. and C.R. Trepte, 1987: Polar stratospheric optical depth observed between 1978 and 1985. *J. Geophys. Res.*, **92**, 4297-4307.

McCormick, M.P., R.E. Veiga and W.P. Chu, 1992: Stratospheric ozone profile and total ozone trends derived from the SAGE I and SAGE II data. *Geophys. Res. Lett.*, **19**, 269-272.

Manabe, S. and R.F. Strickler, 1964: Thermal equilibrium of an atmosphere with a convective adjustment. *J. Atmos. Sci.*, **21**, 361-385.

Miller, A.J., R.M. Nagatani, G.C. Tiao, X.F. Niu, G.C. Reinsel, D.J. Wuebbles and K. Grant, 1992: Comparisons of observed ozone and temperature trends in the lower stratosphere. *Geophys. Res. Lett.* (Submitted).

Penkett, S.A., 1988: Indications and causes of ozone increases in the troposphere. In: *The Changing Atmosphere*. F.S. Rowland and I.S.A. Isaksen (Eds.), Wiley-Interscience, New York, pp91-103.

Prinn, R., D. Cunnold, P. Simmonds, F. Alyea, R. Boldi, A. Crawford, P. Fraser, D. Gutzler, D. Hartley, R. Rosen and R. Rasmussen, 1992: Global average concentration and trend for hydroxyl radicals deduced from ALE/GAGE trichloroethane (methyl chloroform) data: 1978-1990. *J. Geophys. Res.* (In press).

Radick, R.R., G.W. Lockwood and S.L. Baliunas, 1990: Stellar activity and brightness variations: A glimpse at the Sun's history. *Science*, **247**, 39-44.

Ramanathan, V. and R.E. Dickinson, 1979: The role of stratospheric ozone in the zonal and seasonal radiative energy balance of the earth-troposphere system. *J. Atmos. Sci.*, **36**, 1084-1104.

Ramanathan, V., R.J. Cicerone, H.B. Singh and J.T. Kiehl, 1985: Trace gas trends and their potential role in climate change. *J. Geophys. Res.*, **90**, 5547-5566.

Ramaswamy, V., M.D. Schwarzkopf and K.P. Shine, 1992: Radiative forcing of climate from halocarbon-induced global stratospheric ozone loss. *Nature*, **355**, 810-812.

Reid, G.C., 1991: Total solar irradiance variations and the global sea surface temperature record. *J. Geophys. Res.,* **96**, 2835-2844.

Rind, D., N.K. Balachandran and R. Suozzo, 1991: Climate change and the middle atmosphere. Part II: The impact of volcanic aerosols. *J. Clim.* (In press).

Rind, D., R. Suozzo, N.K. Balachandran and M.J. Prather, 1990: Climate change and the middle atmosphere. Part I: The doubled CO_2 climate. *J. Atmos. Sci.*, **47**, 475-494.

Shine, K.P., 1987: The middle atmosphere in the absence of dynamical heat fluxes. *Quart. J. Roy. Meteor. Soc.*, **113**, 603-633.

Siegenthaler, U., 1983: Uptake of excess CO_2 by an outcrop-diffusion model of the ocean. *J. Geophys. Res.*, **88**, 3599-3608.

Siegenthaler, U. and H. Oeschger, 1987: Biospheric CO_2 emissions during the past 200 years reconstructed by deconvolution of ice core data. *Tellus*, **39B**, 140-154.

Staehelin, J. and W. Schmid, 1991: Trend analysis of tropospheric ozone concentrations utilizing the 20-year data set of ozone balloon soundings over Payerne (Switzerland). *Atmos. Envir.*, **25a**,1739-1749.

Stolarski, R.S., P. Bloomfield, R.D. McPeters and J.R. Herman, 1991: Total ozone trends deduced from NIMBUS 7 TOMS data. *Geophys. Res. Lett.*, **18**, 1015-1018.

Sze, N.D., 1977: Anthropogenic CO emissions: implications for the atmospheric $CO-OH-CH_4$ cycle. *Science*, **195**, 673-675.

Wang, W.-C. and N.D. Sze, 1980: Coupled effects of atmospheric N_2O and O_3 on Earth's climate. *Nature*, **286**, 589-590.

Wang, W.-C., M.P. Dudek, X.-Z. Liang and J.T. Kiehl, 1991: Inadequacy of effective CO_2 as a proxy in simulating the greenhouse effect of other radiatively active gases. *Nature*, **350**, 573-577.

Wigley, T.M.L., 1991: Could reducing fossil-fuel emissions cause global warming? *Nature,* **349**, 503-506.

Wigley, T.M.L., M. Hulme and T. Holt, 1990: *An alternative approach to calculating global warming potentials.* Presented to a Workshop on the Scientific Basis of Global Warming Potential Indices, Boulder, CO, USA, Nov. 1990.

Wigley, T.M.L. and P.M. Kelly, 1990: Holocene climatic change, ^{14}C wiggles and variations in solar irradiance. *Phil. Trans. Roy. Soc., A330*, 547-560.

Willson, R.C. and H.S. Hudson, 1991: The Sun's luminosity over a complete solar cycle. *Nature*, **351**, 42-44.

WMO, 1986: Atmospheric Ozone 1985. WMO Global Ozone Research and Monitoring Project, Report No. 16, Geneva.

WMO, 1992: *Scientific Assessment of Ozone Depletion.* WMO/UNEP, WMO Global Ozone Research and Monitoring Project, Report No. 25, Geneva.

A3

Emissions Scenarios for the IPCC:
an Update

J. LEGGETT, W.J. PEPPER, R.J. SWART

Contributors:
J.A. Edmonds; L.G. Meira Filho; I. Mintzer; M.-X. Wang; J. Wasson

CONTENTS

EXECUTIVE SUMMARY

Scenarios of net greenhouse gas and aerosol precursor emissions for the next 100 years or more are necessary to support study of potential anthropogenic impacts on the climate system. The scenarios both provide inputs to climate models and also assist in assessing the relative importance of relevant trace gases and aerosol precursors in changing atmospheric composition and hence climate. Scenarios can also help to improve the understanding of key relationships among factors that drive future emissions.

Scenarios are not predictions of the future and should not be used as such. This becomes increasingly true as the time horizon increases, because the basis for the underlying assumptions becomes increasingly speculative. Considerable uncertainties surround the evolution of the types and levels of human activities (including economic growth and structure), technological advances, and human responses to possible environmental, economic and institutional constraints.

Since completion of the 1990 Scenario A (SA90), events and new information have emerged which relate to that scenario's underlying assumptions. These developments include: the London Amendments to the Montreal Protocol; revision of population forecasts by the World Bank and the United Nations; publication of the IPCC Energy and Industry Sub-group scenario of greenhouse gas emissions to AD 2025; political events and economic changes in the former USSR, Eastern Europe and the Middle East; re-estimation of sources and sinks of greenhouse gases (reviewed in this Assessment); revision of preliminary FAO data on tropical deforestation; and new scientific studies on forest biomass.

These factors have led to an update of SA90, the current exercise providing an interim view and laying a basis for a more complete study of future emissions. Six alternative IPCC scenarios (IS92a-f) now embody a wide array of assumptions affecting how future greenhouse gas emissions might evolve in the absence of climate policies beyond those already adopted. The different worlds which the new scenarios imply, in terms of economic, social and environmental conditions, vary widely and the resulting range of possible greenhouse gas futures spans almost an order of magnitude. Overall, the scenarios indicate that greenhouse gas emissions could rise substantially over the coming century in the absence of new and explicit control measures. IS92a is closer to SA90 due to modest and largely offsetting changes in the underlying assumptions. The highest greenhouse gas levels result from IS92e which combines, among

other assumptions, moderate population growth, high economic growth, high fossil fuel availability and eventual hypothetical phase-out of nuclear power. At the other extreme, IS92c has a CO_2 emission path which eventually falls below its 1990 starting level. It assumes that population grows, then declines by the middle of the next century, that economic growth is low, and that there are severe constraints on fossil fuel supplies. IS92b, a modification of IS92a, suggests that current commitments by many OECD Member countries to stabilize or reduce CO_2 might have a small impact on greenhouse gas emissions over the next few decades, but would not offset the substantial growth in the rest of the world. IS92b does not take into account the possibility that such commitments could accelerate development and diffusion of low greenhouse gas technologies, nor possible resulting shifts in industrial mix.

Population and economic growth, structural changes in economies, energy prices, technological advance, fossil fuel supplies, nuclear and renewable energy availability are among the factors which could exert a major influence on future levels of CO_2 emissions. Developments such as those in the republics of the former Soviet Union and in Eastern Europe, now incorporated into all the scenarios, have important implications for future fossil fuel carbon emissions, by affecting the levels of economic activities and the efficiency of energy production and use. Biotic carbon emissions in the early decades of the scenarios are higher than SA90, reflecting higher preliminary FAO estimates of current rates of tropical deforestation in many - though not all - parts of the world, and higher estimates of forest biomass.

The revised scenarios for CFCs and other substances which deplete stratospheric ozone are much lower than in SA90. This is consistent with wide participation in controls under the 1990 London Amendments to the Montreal Protocol. However, the future production and composition of CFC substitutes (HCFCs and HFCs) could significantly affect the levels of radiative forcing from these compounds.

The distribution of CH_4 and N_2O emissions from their respective sources has changed from the SA90 case. CH_4 emissions from rice paddies are lower, and emissions from animal waste and biomass burning have also been revised downwards. Adipic and nitric acid production have been included as additional sources of N_2O. Preliminary analysis of the emissions of volatile organic compounds and sulphur dioxide suggests that the global emissions of these substances are likely to grow substantially in the coming century.

A3.1 Introduction and Background

In January 1989, the Response Strategies Working Group (RSWG) of the Intergovernmental Panel on Climate Change (IPCC) requested a United States/Netherlands expert group to prepare a set of scenarios of global emissions of carbon dioxide (CO_2), methane (CH_4), nitrous oxide (N_2O), halocarbons and the tropospheric ozone precursors nitrogen oxide (NO_x) and carbon monoxide (CO) (IPCC, 1991a). These scenarios were completed in December 1989 for use by the IPCC Science Working Group (WGI) in its assessment of future climate change (IPCC, 1990a and b). New information has become available since the development of the original scenarios. Consequently, in March 1991, the IPCC requested an update of the existing scenarios in light of recent developments and newly adopted policies. The new IPCC mandate explicitly excluded development of new climate policy scenarios (Swart *et al.*, 1991).

This Section first briefly compares the major assumptions on population and economic growth with historical data or other published forecasts. It then summarizes how the original no-climate-policy Scenario A (SA90) has been updated with information which has become available in 1990 and 1991. This produces the new IPCC Scenarios "IS92a" and "IS92b". Because of substantial uncertainty in how the future will evolve, this Section also includes a preliminary assessment of a range of additional no-climate-policy scenarios, as well as comparisons with other published greenhouse gas scenarios. The updated scenarios are set against two studies of the probability distributions of possible CO_2 scenarios, and then are compared to other published "central tendency" scenarios that extend past the year 2000 and up to the year 2100. Finally, we present sector-by-sector discussions of the particular methods and assumptions in the update, along with additional sectoral scenarios. Detailed documentation of the scenarios is available in a supporting document which also provides tabulations of key variables including population, GNP, primary energy consumption, and emissions (Pepper *et al.*, 1992).

A3.2 New Aspects of this Update

The following changes have occurred since the original IPCC scenarios were developed:

- revised World Bank and United Nations (UN) population forecasts;
- a new greenhouse gas scenario from the IPCC Energy and Industry Sub-Group (EIS);
- important political reforms in the USSR, Eastern Europe and other countries, and a war in the Persian Gulf;

- more optimistic assessments of the economical availability of renewable energy resources;
- revised estimates of current sources and sinks of greenhouse gases, reported in Section A1 of this report;
- revised Food and Agriculture Organization (FAO) data on rates of tropical deforestation and several new studies on forest biomass content; and
- the London Amendments to the Montreal Protocol as well as developments in participation and compliance with the Protocol entered into force.

For completeness, several small sources of greenhouse gases are added to this assessment which were not included in the 1990 IPCC Assessment Report. These include N_2O from the production of nitric acid and adipic acid, and CH_4 from animal wastes and domestic sewage (see Sections A1.3 and A1.4). In addition, scientific evidence described in Section A2 underscores the need to consider the full range of gases which influence climate, directly or indirectly. Cognizant of the importance of a comprehensive approach, this Section also provides preliminary estimates of present and future emissions of volatile organic compounds (VOC) and sulphur oxides (SO_x). Their indirect influences on radiative forcing are more uncertain than those of the direct greenhouse gases and it is not yet possible to quantify on an equivalent basis all the direct and indirect forcing. Improved information with which both to estimate the emissions of these gases and to summarize their effects is needed for future assessments.

The reader should be cautioned, however, that none of the scenarios depicted in this section predicts the future. Long-term scenarios provide inputs to climate models and assist in the examination of the relative importance of relevant trace gases, aerosols, and precursors in changing atmospheric composition and climate. Scenarios can also help improve understanding of key relationships among factors that drive future emissions. Scenarios illustrate the emissions which could be associated with an array of possible assumptions regarding demographics, economics, and technological advance. They can help policymakers to consider the directions in which future emissions may evolve in the absence of new greenhouse gas reduction efforts, and the types of change in important parameters which could or would have to occur in order to significantly change future emission paths.

The results of scenarios can vary considerably from actual outcomes even over short time horizons. Confidence in scenario outputs decreases as the time horizon increases, because the basis for the underlying assumptions becomes increasingly speculative. Uncertainties are of two types:

(i) large uncertainty associated with the evolution of future patterns of human activity, such as economic growth and structure, technological advances, or responses to environmental, economic, or institutional constraints; and

(ii) inadequacy of scientific knowledge concerning physical parameters, such as emission factors, and their relationships.

We have addressed some of these uncertainties in the updated IS92a and b in four ways. The first is through the creation of new scenarios (c through f explained in Table A3.1) using modified key parameters; the second compares some of the new IPCC scenarios with two studies which have mapped probability distributions of future CO_2 emissions; the third compares the range of new IPCC scenarios with other published studies of "central tendencies"; and the fourth analyses sensitivity of results in different sectors to key parameters. The current exercise lays a basis for more complete analysis of the credible range and probabilities of alternative scenarios.

A possibly important limitation of this analysis is that it does not assess the effects that climate change may have on agricultural production, energy demand, and terrestrial ecosystems. Nor does it make assumptions about growth of vegetation if CO_2 increases fertilization, or about losses of the forest uptake of CO_2 if deforestation continues. There could also be positive feedbacks on CO_2 and methane emissions through increased respiration of vegetation and degradation of organic soils. Also, as discussed in Section A2 of this report, we do not yet have an adequate method for summarizing on an equivalent basis the effects on climate of all the greenhouse gases.

A3.3 The Analytical Tool: The "Atmospheric Stabilization Framework"

The Atmospheric Stabilization Framework (ASF), developed by the US Environmental Protection Agency (EPA, 1990), was used as the primary tool for integrating the assumptions and estimating future emissions of greenhouse gases. The ASF is a framework which combines emission modules for various sectors including modules for energy, industry, agriculture, forests and land conversion, as well as a number of small sources. Each module combines assumptions concerning population, economic growth, structural change, resource availability, and emission coefficients to estimate emissions in future time periods. The ASF estimates emissions for CO_2, CH_4, N_2O, chlorofluorocarbons (CFCs) and substitutes, CO, NO_x, VOCs, and SO_x. The energy module uses energy prices to equilibrate supply from the different energy supply sources with demand in four energy end-use sectors. Increased energy prices encourage additional energy supply and increases in energy efficiency but have only a small feedback on economic growth. The agriculture module combines assumptions on population

growth, economic growth, improved yields, and other factors to estimate future production and consumption of agricultural products, land use, and fertilizer use along with emissions of greenhouse gases from these activities. The CFC module estimates future emissions of CFCs, HCFCs, carbon tetrachloride (CCl_4), methyl chloroform, and HFCs under different policy objectives and compliance scenarios, derived with assumptions on the estimated growth in demand for CFCs. The tropical forest section combines assumptions including population, demand for agricultural land, method of forest clearing, and the amount of biomass stored in the vegetation and soils to estimate the clearing of tropical forests and the fate of the cleared land (e.g., forests may be cleared and then allowed to lie fallow in which case they can start to re-accumulate carbon).

A3.4 General Assumptions

The assumptions for the scenarios in this report come mostly from the published forecasts of major international organizations or from published expert analyses. Most of these have been subject to extensive review. The premises for the 1992 IPCC Scenarios a and b ("IS92a" and "IS92b") most closely update the SA90 scenario from IPCC (1990) There exist a wide variety of other plausible assumptions, some of which are used in the range of new scenarios presented here. Table A3.1 summarizes the different assumptions used in the six scenarios. These assumptions are documented in detail in a supporting report (Pepper *et al.*, 1992).

New information since the SA90 has raised the assumed population and economic growth rates compared to those in the earlier assessment. The forecast of future population growth in the SA90 came from the World Bank (Zachariah and Vu, 1988), which the World Bank has since revised (Bulatao *et al*, 1989). The UN has also published new population estimates (UN, 1990; UN Population Division, 1992). The UN medium case is very close to the World Bank's update; the UN medium-low and medium-high cases are used in the alternative scenarios presented here. Most of the variance between the medium-high and medium-low cases is in the developing countries. The updated World Bank population assumptions are close to 10% higher than the assumptions in the SA90: global population increases from 4.84 billion in 1985, to 8.42 billion in 2025, and to 11.33 billion in 2100, with about 94% of the growth occurring in the developing countries. The UN medium forecast of future population estimates that global population may reach 8.51 billion in 2025, 1% higher than the World Bank estimate. Their recent medium extension of this suggests world population of 11.18 billion in 2100 (UN Population Division, 1992), or about 1% lower than the World Bank's. The UN medium-low

Table A3.1: *Summary of assumptions in the six IPCC 1992 alternative scenarios.*

Scenario	Population	Economic Growth		Energy Supplies	Other	CFCs
IS92a	World Bank 1991 11.3 B by 2100	1990-2025: 1990-2100:	2.9% 2.3%	12,000 EJ Conventional Oil 13,000 EJ Natural Gas Solar costs fall to $0.075/kWh 191 EJ of biofuels available at $70/barrel †	Legally enacted and internationally agreed controls on SO_x, NO_x and NMVOC emissions. Efforts to reduce emissions of SO_x, NO_x and CO in developing countries by middle of next century.	Partial compliance with Montreal Protocol. Technological transfer results in gradual phase out of CFCs in non-signatory countries by 2075.
IS92b	World Bank 1991 11.3 B by 2100	1990-2025 1990-2100	2.9% 2.3%	Same as "a"	Same as "a" plus commitments by many OECD countries to stabilize or reduce CO_2 emissions.	Global compliance with scheduled phase out of Montreal Protocol.
IS92c	UN Medium-Low Case 6.4 B by 2100	1990-2025 1990-2100	2.0% 1.2%	8,000 EJ Conventional Oil 7,300 EJ Natural Gas Nuclear costs decline by 0.4% annually	Same as "a"	Same as "a"
IS92d	UN Medium-Low Case 6.4 B by 2100	1990-2025 1990-2100	2.7% 2.0%	Oil and gas same as "c" Solar costs fall to $0.065/kWh 272 EJ of biofuels available at $50/barrel	Emission controls extended worldwide for CO, NO_x, NMVOC and SO_x. Halt deforestation. Capture and use of emissions from coal mining and gas production and use.	CFC production phase out by 1997 for industrialized countries. Phase out of HCFCs.
IS92e	World Bank 1991 11.3 B by 2100	1990-2025 1990-2100	3.5% 3.0%	18,400 EJ conventional oil Gas same as "a" Phase out nuclear by 2075	Emission controls which increase fossil energy costs by 30%.	Same as "d"
IS92f	UN Medium-High Case 17.6 B by 2100	1990-2025: 1990-2100:	2.9% 2.3%	Oil and gas same as "e" Solar costs fall to $0.083/kWh Nuclear costs increase to $0.09/kWh	Same as "a"	Same as "a"

† - Approximate conversion factor: 1 barrel = 6GJ.

Table A3.2: *Population assumptions (in millions).*

Region	World Bank (1991) IS92a, b & e			UN Med-Low IS92c & d		UN Med-High IS92f	
	1990	2025	2100	2025	2100 [†]	2025	2100 [†]
OECD	838	939	903	865	503	1,039	1,359
USSR & E.-Europe	428	496	513	475	337	540	856
China & CP Asia [††]	1,218	1,721	1,924	1,526	935	1,881	2,385
Middle East	128	327	603	272	223	349	693
Africa	648	1,587	2,962	1,375	1,668	1,807	4,651
Latin America	440	708	869	682	770	832	1,662
South & East Asia	1,553	2,636	3,538	2,395	1,979	2,999	5,987
Total	5,252	8,414	11,312	7,591	6,415	9,445	17,592

[†]   Regional breakout of data for 2100 reported here was derived from UN Population Division (1992) which used different regions from this IPCC analysis and did not provide country-specific estimates.

[††]   CP = Centrally planned economies.

Table A3.3: *GNP growth assumptions (average annual rate)*

	.... World Bank [†]			IS92c		IS92a		IS92e	
	Low [††]	High [††]							
	1965	1990	1990	1990	1990	1990	1990	1990	1990
	1989	2000	2000	2025	2100	2025	2100	2025	2100
OECD	3.2%	2.4%	3.1%	1.8%	0.6%	2.5%	1.7%	3.0%	2.2%
USSR/E.Europe [†††]	1.3%	3.2%	3.6%	1.5%	0.5%	2.4%	1.6%	3.2%	2.4%
China & CP [¥] Asia	7.6%	5.6%	6.7%	4.2%	2.5%	5.3%	3.9%	6.1%	4.7%
Other	4.7%	3.8%	4.5%	3.0%	2.1%	4.1%	3.3%	4.8%	4.1%
Global	-	-	-	2.0%	1.2%	2.9%	2.3%	3.5%	3.0%

[†]   Source: World Bank (1991).

[††]   Estimated using projections of regional growth in GDP/capita and country estimates of GDP, population and population growth from 1990 to 2000.

[†††]   World Bank data only include several countries in Eastern Europe.

[¥]   CP = Centrally planned economies.

and medium-high cases reach 6.4 billion and 17.6 billion, respectively, in 2100. Table A3.2 summarizes the population assumptions by region.

Future economic growth assumptions are summarized in Table A3.3. Growth rates assumed in this update are based in part on the reference scenario to 2025 of the Energy and Industry Sub-Group (EIS) of the RSWG (IPCC, 1991b). However, we have adjusted them downward in the near and medium terms in Eastern Europe, the (former) USSR, and the Persian Gulf due to the likely impacts of recent political events. Overall, the economic growth assumptions in IS92a and IS92b are higher than those used in the SA90, especially in Africa, China and Southeast Asia. However,

the IS92a and b assumptions for 1990 to 2000 are generally below or at the low end of the ranges forecast by the World Bank (1991), with the exceptions of the US, OECD (Organization of Economic Cooperation and Development) Pacific and the Middle East. Table A3.3 compares the ranges of GNP assumptions in the new IPCC Scenarios with historical data and the near term projections from the World Bank (World Bank, 1991). The GNP growth rate assumptions for the initial 35 years of IS92a and b, from 1990 to 2025, are substantially below those experienced by most world regions in the past 34 years, from 1955 to 1989. The exceptions are Africa; and Eastern Europe and the USSR, where we assume substantial

Table A3.4: *GDP per capita growth † from 1955 to 1989 (average annual rate)*

OECD Members	2.9%
Eastern Europe	2.5%
Europe, Middle East, & North Africa	2.6%
Sub-Saharan Africa	1.0%
Latin America	1.6%
Asia	4.6%

†   Source: World Bank (1991), rates of growth re-estimated.

Table A3.5: *GNP per capita growth assumptions from 1990 to 2025 (average annual rate)*

	IS92c	IS92a	IS92e
OECD	1.7%	2.2%	2.7%
USSR/E.-Europe	1.2%	2.0%	2.7%
China & CP Asia	3.5%	4.3%	5.1%
Mid-East	0.5%	1.2%	2.0%
Africa	0.9%	1.7%	2.5%
Latin America	1.2%	1.9%	2.7%
Rest of Asia	2.3%	3.0%	3.8%

structural adjustment would boost growth in the medium term. Over the long-term, GNP tends to slow due to an expected slowing of population growth.

Income per capita is assumed to rise most rapidly in the developing world throughout the next century, but in 2100 it remains well below levels in the developed economies. Table A3.4 provides historical data from the World Bank on growth of GDP/capita in the past 34 years, for comparison with Table A3.5 showing the range of IS92 assumptions for 1990 to 2025.

It is important to note that the emission results provided in this paper are highly sensitive to both the population and economic growth assumptions. These parameters would most likely be negatively correlated, however. The economic growth assumptions in IS92a and b fall below historical rates. It is uncertain whether ambitious growth can be realized and maintained in all regions, especially considering possible capital and resource constraints and the presently volatile circumstances in a number of nations. On the other hand, the relatively low GNP per capita in developing countries even at the end of the period suggest that the IS92a and b assumptions fall well below the aspirations of many countries.

All the scenarios presented here include the changes in

government policies aimed at mitigating climate change which have been adopted (as of December 1991). The expert group used a rule for IS92a to incorporate only those emission controls internationally agreed upon and national policies enacted into law, such as the London Amendments to the Montreal Protocol, the amended US Clean Air Act, and the SO_x, NO_x, and VOC Protocols of the Convention on Long Range Transboundary Air Pollution (LRTAP).[1] It does not include the CO_2 emissions targets that have been proposed but not enacted by many OECD countries nor broad policy proposals to cut back on deforestation.

Another scenario, IS92b, shows some of the uncertainty surrounding these policies. IS92b enlarges the interpretation of current policies to include stated policies beyond those legally adopted. All CO_2 commitments of OECD countries, for example, are included, along with an assumption of worldwide ratification and compliance with the amended Montreal Protocol. Since we assume that the CO_2 stabilization commitments would be achieved through improvements in energy efficiency and switching from fossil fuels to nuclear or renewable energy, this reduces simultaneously some of the other greenhouse gases emitted by fossil fuels. We assume that after the target years, emissions would be kept level.

Four additional scenarios have been constructed to examine the sensitivity of future greenhouse gas emissions to a wider range of alternative input assumptions for key variables. Full documentation is available in Pepper *et al.* (1992). The scenarios suggest very different pictures of the future.

IS92c, the lowest scenario, assumes the UN medium-low population forecast, in which population declines in the twenty-first century. It also assumes lower growth in GNP per capita than IS92a and b, as well as low oil and gas resource availability, resulting in higher prices and promoting expansion of nuclear and renewable energy. Deforestation would be slower with lower population growth. IS92d represents another low but more optimistic scenario. It extrapolates some possible trends towards increasing environmental protection, but includes only actions that could be taken due to concern about local or regional air pollution, waste disposal, etc. Population growth assumes the UN medium-low forecast and would be associated with lower natality, falling below the replacement rate late in the twenty-first century, due for example to improvement in per capita income or increased

---

1- In addition, since there is a trend toward increasing control of local air pollution in many countries, all scenarios assumed a slow penetration of low-cost technologies which reduce SO_x and NO_x emissions from very large installations and, in the case of NO_x and CO, from motor vehicles.

Table A3.6: *Selected results of six 1992 IPCC greenhouse gas scenarios.*

| Scenario | Years | Decline in TPER/GNP (AARC) | Decline in C Intensity (AARC) | Cumulative Net Fossil C Emissions (GtC) | Tropical Deforestation | | Year | Emissions Per Year | | | | | | | |
					Total Forest Cleared (million hectares)	Cumulative Net C Emissions (GtC)		CO₂ (GtC)	CH₄ (Tg)	N₂O (TgN)	CFCs (kt)	HCFCs (kt)	HFCs (kt)	Other halo-carbons (kt)	SOₓ (TgS)
IS92a	1990-2025 1990-2100	0.8% 1.0%	0.4% 0.2%	285 1386	678 1447	42 77	1990 2025 2100	7.4 12.2 20.3	506 659 917	12.9 15.8 17.0	827 217 3	143 824 1074	0 511 1823	864 121 0	98 141 169
IS92b	1990-2025 1990-2100	0.9% 1.0%	0.4% 0.2%	275 1316	678 1447	42 77	2025 2100	11.8 19.1	659 917	15.7 16.9	36 0	847 1075	533 1823	3 0	140 164
IS92c	1990-2025 1990-2100	0.6% 0.7%	0.7% 0.6%	228 672	675 1343	42 70	2025 2100	8.8 4.6	589 546	15.0 13.7	217 3	824 1074	511 1823	121 0	115 77
IS92d	1990-2025 1990-2100	0.8% 0.8%	0.9% 0.7%	249 908	420 651	25 30	2025 2100	9.3 10.3	584 567	15.1 14.5	24 0	316 0	1064 2764	3 0	104 87
IS92e	1990-2025 1990-2100	1.0% 1.1%	0.2% 0.2%	330 2050	678 1447	42 77	2025 2100	15.1 35.8	692 1072	16.3 19.1	24 0	316 0	1064 2764	3 0	163 254
IS92f	1990-2025 1990-2100	0.8% 1.0%	0.1% 0.1%	311 1690	725 1686	46 93	2025 2100	14.4 26.6	697 1168	16.2 19.0	217 3	824 1074	511 1823	121 0	151 204

TPER = Total Primary Energy Requirement
Carbon (C) intensity is defined as units of carbon per unit of TPER
AARC = Average annual rate of change
CFCs include CFC-11, CFC-12, CFC-113, CFC-114 and CFC-115
Other halocarbons include carbon tetrachloride, methyl chloroform, Halon 1211 and Halon 1301

family planning. Like IS92c, low fossil resource availability gives rise to greater market penetration of renewable energy and safe nuclear power. The costs of more stringent local pollution controls are incorporated into IS92d through a 30% environmental surcharge on fossil energy use. Greater well-being is assumed to lead to voluntary actions to halt deforestation, to adopt CFC substitutes with no radiative or other adverse effects, and to recover and efficiently use the CH_4 from coal mines and landfills.

IS92e, the case with the highest estimated CO_2 emissions, assumes the World Bank (moderate) population forecast but a more rapid improvement in GNP per capita. Fossil resources are plentiful, but, due to assumed improvement in living standards, environmental surcharges are imposed on their use. Nuclear energy is phased out by 2075. CFC substitute assumptions are like those in IS92d, but plentiful fossil resources discourage the additional use of coal mine methane for energy supply as assumed in IS92d. Deforestation proceeds at the same pace

as in IS92a. IS92f falls below IS92e, using the high UN population growth forecasts but the lower assumptions of improvement in GNP per capita than IS92a. Other assumptions are high fossil resource availability, increasing costs of nuclear power, and less improvement in renewable energy technologies and costs. Table A3.6 summarizes key results of the scenarios for all gases and Figure A3.1 illustrates the scenarios for CO_2 emissions only. Table A3.7 summarizes net emissions of CO_2, and anthropogenic emissions of CH_4, and N_2O by region for IS92a.

A3.5 Comparisons with Other Studies

No systematic analysis has been conducted in this exercise of the likelihood of any of the outcomes illustrated in the six new IPCC Scenarios described above. However, the probabilities of these emission paths has been considered in part by comparison with other studies of probabilities

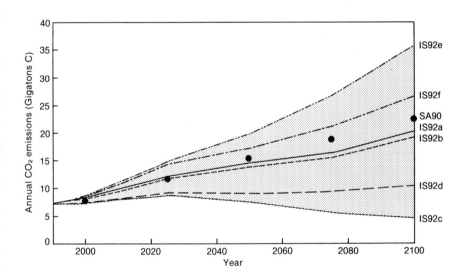

Figure A3.1: Annual CO_2 emissions from energy, cement production and tropical deforestation for the six IPCC 1992 scenarios (IS92a-f) and for the 1990 IPCC Scenario A (SA90).

Table A3.7: Net emissions of CO_2 and anthropogenic emissions of CH_4 and N_2O under IS92a.

	CO_2 (GtC)			Anthropogenic N_2O (TgN)			Anthropogenic CH_4[†] (Tg)		
	1990	2025	2100	1990	2025	2100	1900	2025	2100
OECD	2.8	3.5	4.3	1.4	2.0	1.7	74	90	143
USSR & Eastern Europe	1.7	2.4	2.5	0.8	1.0	0.9	70	62	113
China and CP Asia[††]	0.6	1.6	4.2	0.5	0.8	0.9	37	53	73
Other	0.9	3.2	8.8	1.9	3.7	5.1	146	260	380
Total	6.0	10.7	19.8	4.6	7.5	8.7	326	465	709

[†]   Excludes CH_4 from domestic sewage which was not estimated on a regional basis.
[††] CP = Centrally planned economies.

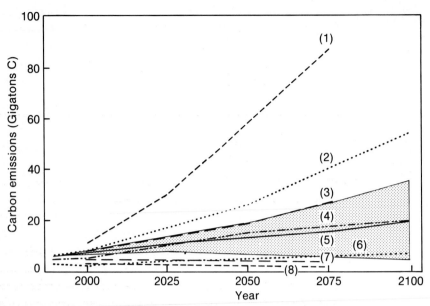

Figure A3.2: Uncertainty analyses for carbon emissions from fossil fuel. The following studies are represented: Edmonds - (1) 95th percentile, (3) 75th percentile, (7) 25th percentile, and (8) 5th percentile; Nordhaus and Yohe - (2) 95th percentile, (4) 50th percentile, and (6) 5th percentile; also (5) IS92a. The shaded area indicates the range of the IS92 scenarios.

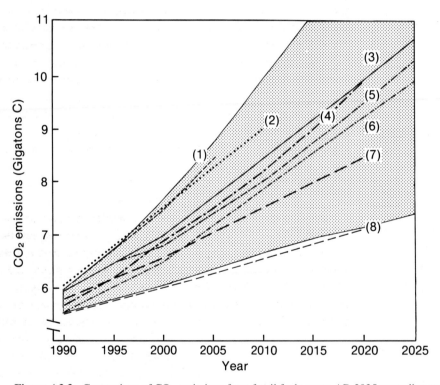

Figure A3.3: Comparison of CO_2 emissions from fossil fuels out to AD 2025 according to: (1) IEA; (2) CEC; (3) IS92a; (4) OECD; (5) IS92b; (6) SA90; (7) the World Energy Conference (WEC) "moderate" and (8) "low" scenarios. The shaded area indicates the range of the IS92 scenarios.

and "best guess" scenarios. Two studies, Edmonds *et al.* (1986) and Nordhaus and Yohe (1983) explicitly examined the issue of uncertainty by estimating the probabilities associated with the various critical input assumptions and the correlation among them, in order to calculate probability distributions of future CO_2 trends. Figure A3.2 shows their results, against a shaded area representing the

range of the new IPCC Scenarios. Both the Edmonds *et al.* and the Nordhaus and Yohe teams found a very wide variation in potential future emissions in the absence of policies to limit this growth. The IS92a falls slightly below the 50th percentile values from Nordhaus and Yohe in the 2050 to 2100 time period until they coincide near 2100. IS92b is slightly lower. The IS92a and b lie between the

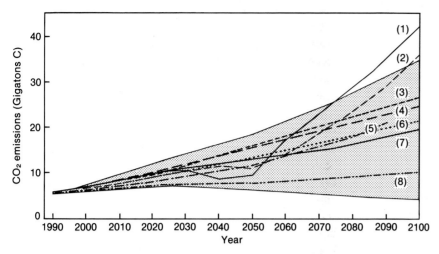

Figure A3.4: Comparison of CO_2 emissions from fossil fuels according to longer-range scenarios: (1) CETA; (2) CRTM-RD; (3) Manne & Richels; (4) EPA (RCW); (5) Edmonds/Reilly; (6) SA90; (7) IS92a, and (8) EPA (SCW). The shaded area indicates the range of the IS92 scenarios.

50th and 75th percentiles in the Edmonds analysis. The highest, new IPCC Scenario is very close to Edmonds 75th percentile, while the lowest is higher than Edmonds 25th percentile. These results from Edmonds *et al.* and Nordhaus and Yohe reinforce the conclusion that emission trajectories are extremely sensitive to a number of key parameters including population growth, economic growth, improvements in energy efficiency and structural change in the economy, as well as the future costs of fossil energy and alternative energy supplies.

Figure A3.3 compares CO_2 emissions from the range of IS92 scenarios in the period 1990 to 2025 to emissions adapted from scenarios of future energy use developed by the International Energy Agency (IEA, 1991), the Commission of the European Communities (CEC, 1989), the World Energy Conference (WEC, 1989), Burniaux *et al.* (1991), and SA90. Emissions in the IEA and CEC Scenarios are from 5% to 12% higher than the emissions in the IS92a through to 2010, the end point of their scenarios. Burniaux *et al.* (1991), using the OECD-GREEN model, suggest CO_2 emissions that are 4% lower than the IS92a in the early years but increase to the same levels by 2020. Emissions in the WEC moderate scenario are 24% lower than IS92a in 2020. Figure A3.4 examines longer-term scenarios, comparing the new IPCC Scenarios with 7 others, including the SA90 Scenario, two scenarios (Slowly Changing World, SCW, and Rapidly Changing World, RCW) developed by the US EPA (1990), scenarios from Manne and Richels (1990), and several studies underway within the Stanford University Energy Modeling Forum. The modellers include Edmonds and Barnes, Peck and Teisberg (using CETA), and Rutherford (using CRTM) (Weyant, 1991: studies to be published in mid-1992). The range of the new IPCC Scenarios is broader than the "central tendency" studies presented in Figures

A3.3 and A3.4. This is especially true by 2100, due to the range of population and economic growth assumptions used in the scenarios. Edmonds *et al.* (1992) provides a more detailed comparison of these scenarios.

A3.6 Energy

Future levels of greenhouse gas emissions from the energy sector are a function primarily of population, incomes, the structure and efficiency of economies, and the relative costs and availability of different sources of energy. The population and economic assumptions in this update of the IPCC scenarios have already been discussed. To 2025, the estimates of energy demand by region and sector are based primarily on the EIS reference scenario. After 2025, energy demand is a function of the economic assumptions and the factors discussed in this Section, and modelled by the ASF (EPA, 1990). Associated emissions of greenhouse gases are estimated using coefficients from the OECD (1991).

The exogenous assumptions of improvements in the intensity of energy end-use are critical parameters counterweighing the upward push on CO_2 emissions from population and economic growth. The assumptions used within IS92a and b result in a global decrease in energy intensity of 0.8% annually in the period to 2025, and 1.0% annually from 2025 to 2100. The decrease in energy per unit of GNP is assumed to be particularly strong in China through to 2100 and in Eastern Europe and the former republics of the Soviet Union in the period 2000 to 2025 as IS92a and b assume substantial structural change. This reflects a complex mix of factors, including market-oriented reforms, a tendency to increase energy demand per capita with increased standards of living (though GNP growth is substantially reduced in Eastern Europe and the republics of the former Soviet Union in the early part of this period), and a tendency toward consumption of less energy intensive goods and services as economies develop. Long-term rates of decrease of exogenous end-use intensity are plausible in the range of 0.8% to 1.0% per year. Edmonds and Reilly (1985) explored the historical rate of change of the energy to GNP ratio in a wide range of OECD countries over extended periods of up to 100 years. The United Kingdom averaged a rate of decline in the energy to GNP ratio of 0.8% annually from 1880 to 1975, while the United States averaged 1.0% from 1920 to 1975. Rates of change over a 50 year period (1925 through 1975) for 13 other OECD nations ranged from 0.45% per year (Germany) to 1.33% per year (Italy). For three centrally planned economies (the former USSR, Czechoslovakia and Yugoslavia) the rate of change was 1.1% per year over the same 50 year period. The range of experience is significantly broader over shorter periods of time. Globally, according to World Bank data, the ratio of primary energy to GDP declined at an average rate of 0.4% in the 24 year period from 1965 to 1989 (World Bank, 1991).

The shares of different primary energy supplies change dramatically in all of the scenarios due to assumed limitations in fossil resources, expected advances in energy technologies, as well as calculated increases in energy prices. In IS92a and b, assumptions regarding available conventional oil, natural gas liquids (NGL), and gas resources are based on Masters *et al.* (1991). At oil and gas prices of $23 per barrel ($/bbl) and $2 per gigajoule ($/GJ), respectively, the available resource equals the Masters *et al.* mean resource estimates. Additional conventional resources, up to the Masters *et al.* 90% bounding values, are available at prices rising to $35/bbl for oil and $8/GJ for gas. As a result, in these scenarios, conventional oil and gas production is gradually replaced by unconventional fossil sources, by synthetic fuels from coal, and non-fossil energy supplies. World oil prices increase to $55/bbl by 2025 and up to $70/bbl by 2100. Up to 760 billion barrels of unconventional liquids including heavy oils, bitumen, and enhanced oil recovery are assumed to become available at oil prices of $35 to $70/bbl. As conventional oil and gas are replaced with coal and unconventional supplies, the production and conversion losses tend to result in significant increases in the rate of CO_2 emissions in IS92a. Large resources of shale oil are assumed to be available at prices over $70/bbl. For the high fossil resource assumptions in IS92e and f, we incorporated estimates of oil resources from Grossling and Nielson (1985).

Simultaneously, the costs of non-fossil energy supplies are assumed to fall significantly over the next hundred years. For example, solar electricity prices are assumed to fall to US$ 0.075 kilowatt per hour ($/kWh) in the IS92a,

and to US$ 0.065/kWh in the more optimistic IS92d. They are assumed to fall only to $0.083/kWh in IS92f. Overall, while renewables are not significantly competitive with fossil energy in 1990, their market penetration speeds as unit costs fall and fossil fuel prices increase rapidly around 2025. By 2100, nuclear, solar, hydropower and biofuels represent 43% of total global primary energy supply in IS92a and b. Overall, the six new IPCC Scenarios represent a global energy system which continues to be dependent on fossil fuels, notably coal, and which shows only moderate gains in energy efficiency and technological development of non-fossil energy sources.

Global energy use rises steadily until 2100 in IS92a, reflecting increases in population and income. Commercial primary energy use which is 344 exajoules (EJ) in 1990, grows to 708 EJ by 2025 and to 1453 EJ by 2100.[1] Primary fossil energy use represents over 85% of primary energy use until 2025. After 2025, the fossil share of primary energy declines to 57% by 2100 despite a nearly three-fold increase in fossil energy production and use. Energy sources such as nuclear, hydropower, solar, and commercial biofuels play a much more important role after 2025 than before in all scenarios, although the mix varies among scenarios.

CO_2 emissions from energy use increase with primary energy use but at a slightly slower rate due to increased use of non-carbon energy sources. CO_2 emissions in the IS92a increase from 6.0 billion tons of carbon (GtC) in 1990 to 10.7 GtC in 2025 and 19.8 GtC in 2100.[2] Compared to the original SA90, this represents faster growth before 2025 but slower growth afterwards. The higher path until 2025 reflects the assumptions in IS92a and b of more rapid population and economic growth, combined with less optimistic assumptions concerning improvements in energy efficiency, based on the EIS reference scenario assumptions. After 2025, IS92a relies more heavily on non-fossil energy sources than SA90 thereby reducing the rate of growth in CO_2 emissions.

While global CO_2 emissions from energy use in IS92a and b grow fairly constantly at a rate of 1.1% and 1.0% annually, respectively, from 1990 to 2100 (see Figure A3.4), the pattern of growth differs considerably among regions. CO_2 emissions from the developed economies grow at an average annual rate of 0.8% until 2025. After 2025 in the developed economies, lower economic growth combined with stabilized population levels and increasing

---

1 - Primary energy use in SA90 was 473 to 657 EJ in 2025 and 728 to 1682 EJ in 2100, the range reflecting alternative economic growth projections.
2 - Average CO_2 emissions from energy in SA90 were 9.9 GtC in 2025 and 21.7 GtC in 2100.

use of non-fossil energy sources yield reductions in this growth to an annual average of 0.2%. Emissions from developing economies continue to rise due to increases in population and economic growth. In the period, 1990 to 2100, CO_2 emissions per capita from energy use grow at an average annual rate of only 0.2% to 0.3% in the OECD, Eastern Europe, and the republics of the former Soviet Union, while averaging 1.0% for the rest of the world. CO_2 emissions per capita in developing countries remain on average one quarter to one half those of developed countries by 2100. Conversely, CO_2 emissions per dollar GNP (per \$GNP) in the OECD are two thirds the global average, one half those of Eastern Europe and the former Soviet Union, and one fifth those of China. Global average CO_2 emissions per \$GNP decline at an average annual rate of 1.2% from 1990 to 2100. Regionally, the highest rate of decline is in China where CO_2 emissions per \$GNP are over four times higher than the global average in 1990.

IS92b incorporates the stabilization goal for fossil carbon dioxide emissions for the year 2000, proposed by many OECD Member countries. If countries achieve these commitments and sustain them through 2100 (which is likely to require programmes beyond those already planned) and the rest of the world does not adopt similar measures, global emissions in 2025 of fossil carbon would be 0.4 GtC lower than in the IS92a. This reduction represents a reduction of 11% in emissions of CO_2 from the OECD from IS92a but only a 4% reduction of global emissions. These results reflect the long-term contribution of the economies of developing countries, the republics of the former Soviet Union, and Eastern Europe to CO_2 emissions.

A3.7 Halocarbons

Halocarbons, including chlorofluorocarbons (CFCs), their substitutes, and other compounds which deplete stratospheric ozone, may have important implications for climate change. Many of these compounds exert a much more powerful direct radiative forcing than CO_2 per molecule. Recently, it has been discovered that the loss of lower-stratospheric ozone can reduce the radiative forcing of the troposphere/surface system, particularly at high latitudes. Hence, ozone depleting molecules can have both positive (direct) and negative (indirect) contributions to radiative forcing. However, the net effect of such halocarbons on globally averaged temperatures or, more broadly, on climate is uncertain at present. As a result, the comparisons of scenarios for these gases are summarized using the kilotons (kt) of the compounds, not the index of direct "Global Warming Potential" (GWP), as calculated in Section A2 of this report.

An important event since the development of the scenarios for the first IPCC assessment is the agreement to adjust and amend the Montreal Protocol in London in 1990 (the "London Amendments"). Most key nations have either now signed the agreement or have pronounced the intention to do so. In IS92a, 70% of the developing world is assumed to ratify and comply with the London Amendments. This percentage is based on the GNP of countries that have signed and/or ratified as of December 1, 1991 (e.g., China has signed while India has not). We further assume in IS92a that if most of the world develops and uses CFC substitutes, then the need to trade in global markets and "technology transfer" will lead to a gradual phase-out of all CFC use (we assume gradually from 2025 to 2075) even without worldwide ratification. We have also included the voluntary reductions ahead of schedule achieved by many countries. IS92b assumes global compliance with the Montreal Protocol.

The London Amendments contain a recommendation only to use halocarbon substitutes for a transition period. There are no international agreements for eventually phasing down the production of all substitute compounds. Some substitutes may not deplete stratospheric ozone but may still contribute to climate change. Therefore, in the scenarios, we assume that the production of substitutes would mimic the growth rate of the underlying controlled compounds which they replace under the phase-out, adjusted for market reductions due to non-chemical substitution and increased use of recycling and other emission control programmes. Accordingly, all of the cases assume that the demand for CFCs grows by 2.5% annually until 2050 then remains flat. HCFCs and HFCs are assumed to replace approximately 21 to 42% (depending on the scenario) of phased-out CFCs. Substitution is weighted much more towards HCFCs than HFCs over the long-term unless additional policy steps are taken.

This analysis includes seven cases of future emissions of CFCs and their substitutes. Three of these cases are incorporated into the IS92 Scenarios. The first two cases, "Partial Compliance and High HCFC" and "Partial Compliance and Reduced HCFC" portray a future where only 70% of the developing world ratifies and complies with the London Amendments. In these scenarios, CFC production in the remaining 30% continue to grow until 2100. Also, the HCFC reductions in the US are not incorporated. The third scenario, "Partial Compliance and Technology Transfer", is incorporated into IPCC Scenarios IS92a, c, and f. It assumes partial compliance with the London Amendments but assumes that "technology transfer" results in a full phase-out of production of CFCs by 2075. It includes the phase-out of HCFCs in the US required by the Clean Air Act. The fourth case, "Global Compliance", is incorporated in IS92b and contains full global ratification and compliance with the London Amendments. The fifth case, "97 Phase-Out for Developed Countries", accelerates the phase-out schedule for CFCs.

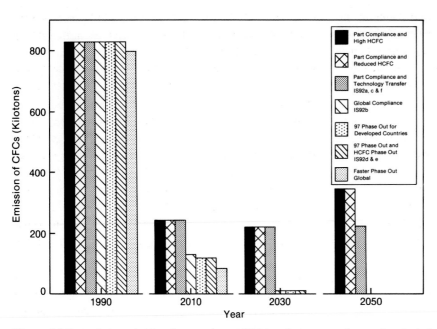

Figure A3.5: Emission of chlorofluorocarbons (CFCs) under a range of scenarios. Includes CFC-11, CFC-12, CFC-113, CFC-114 and CFC-115.

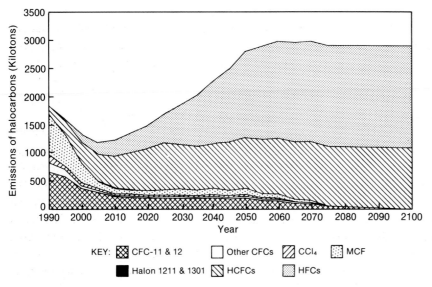

Figure A3.6: Emission of halocarbons under IS92a, c and f. These assume only partial compliance with the London Amendments to the Montreal Protocol but, through technical transfer to non-complying countries, a complete end to CFC production by AD 2075.

The sixth case, "97 Phase-Out and HCFC Phase-Out" is incorporated in IS92c and d and expands on the fifth case by incorporating a global phase-out of HCFCs. The seventh case, "Faster Phase-Out - Global", accelerates the phase-out of CFCs and HCFCs in developing countries.

For the calculations, US EPA's Integrated Assessment Model for CFCs was used. Detailed results can be found in the supporting document, Pepper *et al.* (1992). In the IS92a, emissions of CFCs, carbon tetrachloride, and methyl chloroform decline rapidly through 2010 (see Figures A3.5 and A3.6, "Partial Compliance and Technology Transfer Case"). After 2010, these emissions

stabilize and ultimately decline to zero after 2075 as all the world adopts the prominent technologies. Emissions of HCFCs and HFCs grow rapidly throughout the whole time horizon in all cases, reflecting their roles as substitutes for the CFCs and the postulation of no controls on their emissions (except on HCFCs in the US). Its results indicate that the composition of substitutes could have an important impact on levels of radiative forcing.

If "technology transfer" to non-signatories of the agreement were not to lead to a phase-out of CFCs, the implications by 2100 could be substantial, leading to emissions almost back up to the level estimated for 2000.

The results for the emissions of CFCs are depicted, as an example, in the "Partial Compliance and High HCFC" case in Figure A3.5. In this case, after 2010, emissions of CFCs, carbon tetrachloride and methyl chloroform stabilize and then start to increase. This reflects growth from non-signatories of the London Amendments.

IS92b assumes full ratification of and compliance with the London Amendments. Emissions are nearly eliminated much earlier ("London Amendments-Global Compliance") than in IS92a. Moreover, recent data showing more severe ozone depletion (WMO, 1992) could lead to a more rapid phase-out of CFCs and halons, carbon tetrachloride, and methyl chloroform ("97 Phase-Out for Developed Countries"). Controls on the use of HCFCs are also possible.

A3.8 Agriculture, Forests and Land Conversion

A3.8.1 Agriculture

Details of the estimation of greenhouse gases from agricultural sources are available in Pepper *et al.* (1992). The distribution of the global emissions of non-CO_2 greenhouse gases among different sources in the base year has been taken from the assessment of the global budgets reported in Section A1 of this document. This distribution is still poorly understood for most gases. In particular, emissions of CH_4 from rice cultivation are highly uncertain but lower than believed in IPCC (1990). Average emission coefficients of 38 grams per square metre per year for land under rice cultivation have been selected for all scenarios based on the CH_4 emission budget of 60 teragrams (Tg). In IS92a and b, the emissions rise gradually from 60Tg in 1990 to 88Tg by 2050, then decline to 84Tg by 2100. The scenarios assume continuing advances in crop yields which average 0.5% annually over the period. Consequently, the growth in emissions is slower than growth in rice production, which more than doubles.

In IS92a and b, CH_4 emissions from enteric fermentation in domestic animals rise from 84Tg in 1990 to close to 200Tg by 2100. This increase reflects a rapid increase in consumption of meat and dairy products and assumes constant emissions per unit of production. Emissions of CH_4 from animal wastes have been added, changing with the levels of meat and dairy production. If meat production per animal were to increase, emissions would be lower. Emissions from animal wastes increase from 26Tg CH_4 in 1990 to 62Tg CH_4 by 2100. It is uncertain whether this growth in the production of meat and dairy products can actually be maintained, taking into account possible land and feed constraints which are not explicitly dealt with in these scenarios. Autonomous developments that affect the emissions from enteric fermentation or animal waste per unit of production, such as those resulting from changed feeding patterns, are not hypothesized either. Both types of factors could change the emission trends of these scenarios.

Emissions of N_2O from fertilized soils in 1990 of 2.2 TgN have been selected as the starting budget, falling within the range of uncertainty of 0.3 to 3.0 TgN reported in Section A1. They increase in proportion to fertilizer use, which more than doubles in IS92a and b. The impact of changing fertilization practices and the dependency of N_2O emissions on local soil types, moisture, agricultural practices, etc., has not been estimated.

A3.8.2 Forests and Land Conversion

Since SA90 was finalized, new data have become available regarding both tropical deforestation rates and the average content of carbon per hectare of above-ground vegetation. Both are higher than the assumptions used in SA90. The estimates of carbon in soils and fluxes of greenhouse gases with changes in land uses remain as in SA90. Neither of the possible effects on CO_2 fluxes due to increased fertilization or respiration, which may be associated with higher CO_2 atmospheric concentrations or temperature

Table A3.8: *Assumptions used in deforestation cases.*

Scenario Used In	Biomass Content	Rate of Deforestation
IS92a, b, & e	Moderate	Moderate (tied to moderate population growth)
IS92c	Moderate	Moderate (tied to low population growth)
IS92d	Moderate	Halt Tropical Deforestation
IS92f	Moderate	Moderate (tied to high population growth)
None	High	Moderate (tied to moderate population growth)
None	Moderate	High
None	High	High
None	Moderate	Halt Tropical Deforestation/Increase Establishment of Plantations

increases, have been incorporated in this analysis.

To incorporate the new data and the uncertainties still surrounding these parameters, eight cases of tropical forest clearing and emissions of greenhouse gases were developed. These eight cases include four cases which were incorporated within the new IPCC Scenarios which assume moderate assumptions of rates of tropical deforestation (except IS92d which has a halt to deforestation) and biomass content of vegetation in these forests. The eight cases also include four sensitivities around the case incorporated in IS92a. The sensitivity cases vary rates of deforestation, rates of establishment of plantations, and assumptions concerning the biomass content of the forests. Table A3.8 summarizes these cases and their assumptions.

As lands convert from one use to another, greenhouse gases can be released or taken up by vegetation and soils, for example by the burning or regrowth of forests or the tilling or amendment of soils. This analysis simulates and tracks the changes of land parcels from one use to another from 1975 to 2100, due to agricultural demand, burning, plantations, etc., and calculates the associated greenhouse gas emissions and uptake over time. As land is cleared, sometimes more than once in the period of analysis, only part of the carbon stored in vegetation and soils is released over an extended period of time. As regrowth occurs on cleared land, carbon is sequestered. We calculate the net balance of carbon from vegetation and soils of all lands estimated to be tropical forests at any time from 1975 through 2100. Assumptions about rates of carbon loss or absorption and other parameters used in this analysis, as well as the case results, are detailed in Pepper *et al.* (1992).

The IS92 Scenarios use the new FAO 1990 Tropical Forest Assessment (FAO, 1991) and the 1988 update of the 1980 Tropical Forest Assessment (FAO, 1988) for its deforestation rates. The new assessment estimates that, on average, 17 million hectares of tropical closed and open forest were cleared annually from 1981 to 1990. The 1988 FAO Tropical Forest Assessment provided estimates of clearing rates for the period 1976 to 1980. While questions have been raised concerning the reliability of these data, they are the best and most recent currently available for the world.

We estimated a constant rate of change in clearing rates over this period such that the average quantities of clearing for 1976 to 1980 and for 1981 to 1990 match those reported in the FAO 1988 and 1991 reports, respectively. Moreover, this estimate is constrained so that forest areas in 1980 equal the quantities given in the more complete and detailed 1980 assessment. This results in calculated clearing rates increasing from 13.2 million hectares in 1980 to 19.3 million hectares in 1990. After 1990, deforestation rates increase in proportion to population, but lagged twenty years and constrained by available forest

area in each country. In the "high deforestation" sensitivity cases, these rates are increased by an additional 1% point per year. In the "halt deforestation" sensitivity case and in IS92d, we assume that rates of deforestation decline, starting in 1990. The IPCC Greenhouse Gas Task Force advised that it should be assumed that all forests not legally protected, including areas which have been classified as non-productive, can be subject to deforestation (IPCC, 1991c).

Forest clearing in IS92a, b, c, e, and f increases to 20 to 23.6 million hectares per year by 2025, depending on population growth, and then declines. In IS92d, clearing declines steadily after 1990 to 0.7 million hectares per year by 2025. In the high deforestation sensitivity cases, tropical forest clearing increases to 28.6 million hectares in 2025 before declining. In all sets of assumptions, available forest resources within each country provide upper bounds on future clearing. In the IS92a, 73% of all tropical forests (1.4 billion hectares), are cleared by 2100. In the high deforestation sensitivity case, this fraction increases to 91%. In IS92a, countries representing 43% of forest clearing in 1980 have (or have nearly) exhausted their forest resources by 2025. By 2050, this fraction increases to 52%.

Other factors which vary in the sensitivity cases include the fate of forest fallow, future rates of plantation establishment, and the carbon stored in the aboveground biomass. The high deforestation sensitivity cases include another possible net source of carbon: permanent clearing of forest fallow. These are areas of logged or abandoned agricultural lands which are regenerating to forest. The high deforestation sensitivity cases assume that up to 10 million hectares of forest fallow are currently being converted to permanent agriculture annually (Houghton, 1991) and that this clearing continues into the future until almost all forest fallow is converted. In IS92a, the establishment of plantations, which FAO (1988) estimates as 1.3 million hectares annually between 1980 and 1985, is assumed to continue with 118 million hectares added between 1990 and 2100. The high deforestation sensitivity cases assume that no new plantations are added after 1990.

Moderate estimates of carbon stored in the biomass are from OECD (1991) which have been adapted from Brown and Lugo (1984), Brown *et al.* (1989), and Brown (1991), and have been estimated using wood volumes. The moderate biomass estimates are used for the IS92 Scenarios. The high biomass estimates, used in the sensitivity analyses, increase the moderate estimates by the percentage corrections cited in Houghton (1991) to reflect uncertainties in measurement techniques and results from other studies utilizing alternative (i.e., destructive sampling) approaches. Table A3.9 summarizes current estimates of biomass contents and those used in all IS92 scenarios.

Table A3.9: *Carbon stored in tropical forests (tons C/hectare).*

| | IS92 a,b [†††] | | | | Carbon Stocks from Houghton (1991) | | | | |
| | | | | | Earlier Estimate [†] | | Recent Estimate [††] | | |
	Closed B-leaf	Closed Conif.	Open B-leaf	Crops	Moist Forest (D/V)[†]	Seasonal Forest (D/V)[†]	Closed Forest (U/L)[††]	Open Forest (D/V)[†]	Crops
Vegetation									
Latin America	76	78	27	5	176/82	158/85	89/73	27/27	5
Asia	97	83	27	5	250/135	150/90	112/60	60/40	5
Africa	117	68	16	5	210/124	160/62	136/111	90/15	5
Soils	100	100	69	-	100	90	-	50	-

[†] For columns labelled (D/V), the first value is based on destructive sampling of biomass and the second value is calculated from estimates of wood volumes.

[††] For columns labelled (U/L), the first value is for undisturbed forests and the second value is for logged forests.

[†††] Source: OECD (1991).

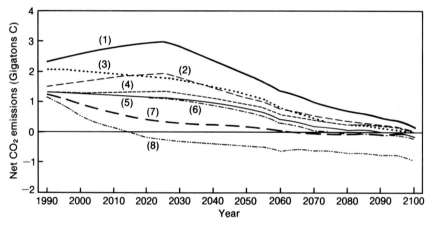

Figure A3.7: Comparison of net CO_2 emissions from tropical deforestation according to some longer-range scenarios: (1) High deforestation/high biomass; (2) High deforestation/moderate biomass; (3) moderate deforestation/high biomass; (4) IS92f; (5) IS92a, b and e - moderate deforestation/moderate biomass; (6) IS92c; (7) IS92d, and (8) No deforestation/high plantation. *(Note: does not include the effects of fertilization or increased respiration on net CO_2 emissions due to higher CO_2 atmospheric concentrations.)*

CO_2 emissions in the IS92 Scenarios and the sensitivity cases range from 1.1 GtC to 2.3 GtC in 1990. These include net soil carbon released as well. In the IS92a, emissions are relatively flat through to 1995 and then start to decline. Net emissions are slightly negative by 2100 due to carbon sequestration by plantations. In the high deforestation and high biomass sensitivity case, emissions increase to 3.0 GtC by 2025, decline to 1.9 GtC in 2050 and 0.2 GtC by 2100. Even though in all of the cases clearing of open forests in the period 1980 to 1990 represents over one third of total clearing, net emissions of CO_2 from open forests represent less than 10% of total deforestation emissions.

All these cases do not span the range of possibilities for both current and future emissions. There is a strong need

for improvement of the base data. Moreover, current trends in deforested area, combined with changes in policies, could very well lead to emissions lower than in the moderate case.

In conclusion, the sensitivity analysis explores a wide range of possible futures and identifies the importance of several key assumptions. Figure A3.7 illustrates the CO_2 emissions from the IS92 Scenarios along with the sensitivities around IS92a. One sensitivity case, "High Deforestation/Moderate Biomass" illustrates the impact of the higher clearing rates and forest fallow clearing assumptions on CO_2 emissions. The sensitivity case, "Moderate Deforestation/High Biomass", illustrates the importance of assumptions concerning carbon stored within the biomass. The sensitivity case, "High Deforestation/High Biomass" shows the combined impact of these alternative assumptions. The "Halt Deforestation/ High Plantations" sensitivity case illustrates the potential for reducing emissions by quickly stopping forest clearing and actively establishing plantations. In this case, forest clearing is reduced starting in 1991 and eliminated by 2025. Up to 293 million hectares of plantations are established by 2100.

A3.8.3 Resolution of Land Assumptions

While these cases and sensitivities provide a plausible range of emissions they do not address all of the uncertainty. A key concern with these cases is consistency of the assumptions about land uses among the different sectors. Because the models and methodologies used for the different sectors are not fully integrated with respect to land use, this consistency is not automatically guaranteed. With population doubling by 2100, improved nutrition in the developing economies, and the agricultural land base possibly degraded, the demand for additional land conversion to produce crops especially in the developing countries would be significant, even if sustained productivity increases were achieved. Above, we already noted the anticipated demand for land for livestock and feed production, while the IS92a assumes over 190 EJ of energy from commercial biofuels by 2100, of which 70% would be in developing countries.

A review of current land uses, forest clearing, and plantation assumptions suggests that the assumptions in the new IPCC Scenarios are not inconsistent. FAO (1986) reported that in 1984, 1.5 billion hectares were used for arable land and permanent crops and 3.1 billion hectares were used for permanent pasture leaving 4.1 billion hectares of forest and woodlands. The moderate defor-estation case assumes that on average 1.6 billion hectares of tropical forests will be cleared by 2100 while 0.1 billion hectares of forest plantations will be established, roughly

allowing for a doubling of the arable land base. Table A3.10 summarizes these statistics for Latin America, Africa, and the Far East along with the IS92a assumptions for these regions. Clearing of forest land in these tropical regions matches well with the increase in population and demand for food. The largest concern would be in Asia where population densities are the highest and cleared forest lands might not have the same productivity as existing cropland, nor may the production be as sustainable. Any reduction in deforestation would reduce the potential land base and would have to be matched by increased productivity on existing agricultural land. Other anthropogenic influences on the terrestrial carbon cycle, such as pollution, erosion, desertification, logging in temperate and boreal zones and carbon sequestration in managed forests, have not been taken into account in this study because of the absence of quantitative information.

A3.9 Other Sectors and Gases

The 1992 Scenarios include emissions of additional gases and emissions from additional sectors. Specifically, emissions of N_2O, CH_4, CO, and NO_x from energy combustion and production were developed using assumptions about emission controls consistent with the US Clean Air Act and the Protocols under the Convention on Long-Range Transboundary Air Pollution. In the base year's emission budget, our estimate of emissions of CO from fossil fuels, by applying emission coefficients to fuel use, falls below the low end of the range given in Section A1 of this report. Our estimate of emissions reflects the latest data on energy consumption and emission coefficients, but the CO budget discrepancies need to be investigated further. Emissions of N_2O from the production of nitric and adipic acid have been included and reflect scenario assumptions on nitrogen fertilizer production and economic growth respectively. Emissions

Table A3.10: *Land-use assumptions (10^6 hectares).*

	Africa	Asia [†]	Latin America
FAO			
1984 Arable Land & Permanent Crops	154	272	177
1984 Permanent Pasture	630	35	551
Remaining Tropical Forest (1980)	704	431	938
Forest Cleared (1980-2100)	653	329	635
Plantation Area Established (1980-2100)	27	47	44
Maximum Energy Plantation Area[††]	88	101	118

[†]   Far East as defined by FAO
[††] Energy plantation area in addition to other plantation area

Table A3.11: *Global emissions of direct greenhouse gases under IS92a.*

	1990	2000	2025	2050	2100
CO$_2$ (GtC)					
Energy	6.0	7.0	10.7	13.2	19.8
Deforestation	1.3	1.3	1.1	0.8	-0.1
Cement	0.2	0.2	0.4	0.5	0.6
Total	7.4	8.4	12.2	14.5	20.3
CH$_4$ (Tg)					
Energy Production & Use	91	94	110	140	222
Enteric Fermentation	84	99	138	173	198
Rice	60	66	78	87	84
Animal Wastes	26	31	43	54	62
Landfills	38	42	63	93	109
Biomass Burning	28	29	32	34	33
Domestic Sewage	25	29	40	47	53
Natural	155	155	155	155	155
Total	506	545	659	785	917
N$_2$O (Tg N)					
Energy	0.4	0.5	0.7	0.8	0.8
Fertilized Soils	2.2	2.7	3.8	4.2	4.5
Land Clearing	0.8	0.8	1.0	1.1	1.0
Adipic Acid	0.5	0.6	0.9	1.1	1.2
Nitric Acid	0.2	0.3	0.4	0.5	0.5
Biomass Burning	0.5	0.6	0.7	0.7	0.7
Natural	8.3	8.3	8.3	8.3	8.3
Total	12.9	13.8	15.8	16.6	17.0
Halocarbons (kilotons)					
CFC-11	298	168	94	85	2
CFC-12	363	200	98	110	1
CFC-113	147	29	21	24	0
CFC-114	13	4	3	3	0
CFC-115	7	5	1	1	0
CCl$_4$	119	34	19	21	0
Methyl chloroform	738	353	97	110	0
HCFC-22	138	275	530	523	614
HCFC-123	0	44	159	214	267
HCFC-124	0	7	11	15	16
HCFC-141b	0	24	82	110	138
HCFC-142b	0	7	11	0	0
HCFC-225	5	17	30	38	40
HFC-134a	0	148	467	918	1055
HFC-125	0	0	14	175	199
HFC-152a	0	0	30	448	570

Table A3.12: Global emissions of indirect greenhouse gases under IS92a.

	1990	2000	2025	2050	2100
CO (TgC)					
Energy	130	140	160	219	385
Biomass Burning	297	309	341	363	353
Plants	43	43	43	43	43
Oceans	17	17	17	17	17
Wildfires	13	13	13	13	13
Total	499	522	574	655	811
NO_x (TgN)					
Energy	25	28	43	53	72
Biomass Burning	9	9	10	11	11
Natural Lands	12	12	12	12	12
Lightning	9	9	9	9	9
Total	55	58	74	85	104
VOC Emissions by Gas (Tg)					
Paraffins	56	59	84	106	152
Olefins	42	43	49	56	72
Aromatics (btx) [†]	15	16	21	25	36
Other	8	8	12	15	23
Total	121	126	166	202	283
VOC Emissions by Sector (Tg)					
Energy Production & Use	27	30	48	64	102
Biomass Burning	53	53	54	54	52
Industry	23	25	36	46	62
Other	18	19	27	38	67
Total	121	126	166	202	283
Sulphur (TgS)					
Energy Production & Use	65	67	101	132	123
Biomass Burning	2	2	3	3	3
Other Industrial	8	10	16	19	21
Natural	22	22	22	22	22
Total	98	101	141	175	169

[†]   Benzene, toulene, and xylene

of CO_2 from cement production and CH_4 from landfills reflect regional population and economic growth assumptions as well as expectations of resource limitations and saturation. Emissions of N_2O, CH_4, CO, and NO_x from biomass burning are based on emission coefficients from Crutzen and Andreae (1990) and Andreae (1991), and future rates of deforestation, clearing of fallow lands for shifting agriculture, non-commercial biofuel use, and other burning activities. Emissions of volatile organic compounds (VOCs) are highly uncertain, especially for developing countries. They are based on a detailed, country level emissions inventory developed by EPA (1991). They are compared with other sources and extended into the future based on activities such as transportation energy use, deforestation and biomass burning, and industrial activity. Emissions of SO_x from energy use are based on a study performed by Spiro *et al.* (1991) where emissions reflect average sulphur content of the different fossil fuels and are adjusted for emission control programmes. Emissions of all trace gases by gas and sector are summarized in Tables A3.11 and A3.12.

All of the new IPCC Scenarios except for IS92d show significant increases in emissions of CO, NO_x and SO_x even the assumed of adoption of some emission controls on large stationary sources and mobile sources in the developing countries. Emissions of these gases could be significantly higher if we had not assumed significant penetration of local pollution controls. For example, in IS92a annual emissions of CO grow from 499 TgC in 1990 to 811 TgC by 2100 and annual emissions of NO_x grow from 55 TgN in 1990 to 104 TgN by 2100. Without the assumed pollution controls, annual emissions of CO and NO_x would increase to 1049 TgN and 108 TgN, respectively, by 2100.

A3.10 Conclusions

This chapter presents a new set of IPCC greenhouse gas emission scenarios. The purpose is not to predict which evolution of greenhouse gases is most probable among the array of plausible alternatives. Rather, comparison of alternatives may help policy-makers to consider the directions in which future emissions may evolve in the absence of new greenhouse gas reduction efforts, and the types of change in important parameters which could or would have to occur to significantly change future paths. Commitments by individual governments or companies to reduce emissions of greenhouse gases in response to the global warming or other environmental issues could significantly affect some of the individual emission sources. However, it is difficult to take these commitments into account in global emission estimates.

Two of the Scenarios "IS92a" and "IS92b" update the original Scenario A from IPCC (1990) by incorporating important information which has become available in 1990 and 1991. IS92a includes only those policies affecting greenhouse gas emissions which are agreed internationally or enacted into national laws (as of December 1991). IS92b includes proposed greenhouse gas policies as well. While some of the revisions to the assumptions used in SA90 are significant, the results of IS92a and b are on balance very similar to the original SA90. The other scenarios provided in this chapter explore a broader range of plausible assumptions than in SA90, and indicate that the array of possible future trends in greenhouse gas emissions spans an order of magnitude. That the different futures within this range are not all equally probable has not been systematically addressed in this analysis, but this topic should be pursued. A more thorough exploration of the uncertainties in assumptions and relationships among

parameters could reveal more about the confidence policymakers should have in such emission scenarios and in the influence of their decisions on future emission paths.

Even with a wide range of possible greenhouse gas scenarios, a number of conclusions can be drawn from the analysis:

- CFC emissions are likely to be substantially lower than previously estimated by the IPCC, especially if technology transfer and world trade requirements lead all countries to comply with the London Amendments to the Montreal Protocol. However, the future production and composition of CFC substitutes could significantly affect the levels of radiative forcing from halocarbons.

- The commitments by many OECD Member countries to achieve and maintain stabilization or reduction of their CO_2 emissions by the year 2000, in absolute terms or per capita, could have an important impact on their own emissions but a small influence on global emissions by the year 2100. The CO_2 commitments may have the simultaneous effect of reducing other greenhouse gases as well. Most of the uncertainty over future growth in greenhouse gas emissions is likely to depend on how developing countries, Eastern European countries, and republics of the former Soviet Union choose to meet their economic and social needs.

- Population growth, robust economic growth, plentiful fossil fuels at relatively low costs, and net deforestation tend to push upward the trends in greenhouse gas emissions; decreases in energy required per unit of GNP, the economical availability of renewable energy supplies, plantations of biomass, the control of conventional air pollution, and improvements in agricultural productivity tend to diminish greenhouse gas emissions over the long-term. Public attitudes and governmental policies may have a strong influence on which of these variables will dominate over the coming century.

- CO_2 emissions from forest and land conversion are higher than previous estimates, ranging from 1.1 GtC to 2.3 GtC in 1990, because of higher recent estimates of both deforestation rates and biomass in forests in the tropics, as assessed by the FAO and several recent studies, respectively.

- There is no evidence to alter the main conclusions of IPCC (1990) regarding CO_2 emissions from fossil fuels. While several factors could lead to both higher or lower emissions, especially in the long-term, most expert analyses suggest that these emissions could rise substantially over the coming century. IS92a and b, which most resemble the SA90 in terms of assumptions for key parameters, show a range of emissions of 10.3 to 10.7 GtC in 2025, and 18.6 to 19.8 GtC in 2100. However, a broader range of alternative assumptions is plausible. The range of emissions estimated for the wider set of alternative scenarios is 7 to 14 GtC in 2025 and 5 to 35 GtC in 2100.

- Comparison with other studies of CO_2 scenarios indicates that the IS92a and b fall within the range of other short-term (to 2025) scenarios, including those of the World Energy Council, the International Energy Agency, and the Commission of the European Communities. By the year 2100, the IS92a and b are below but not distant from all the other long-term CO_2 scenarios reported except one. However, the range of alternative, new IS92 Scenarios encompasses virtually all the other scenarios, with a spread of almost an order of magnitude.

- There remains an important need to improve the estimates of greenhouse gas emissions in all sectors (especially in the forest and agriculture sectors) as well as of the underlying human-related variables (such as rates of land clearing).

References

Andreae, M.O., 1991: Biomass burning: its history, use and distribution and its impact on environmental quality and global climate. In: *Global Biomass Burning, Atmospheric, Climatic, and Biospheric Implications.* J.S. Levine (Ed.). MIT Press, Cambridge, MA.

Brown, S. and A.E. Lugo, 1984: Biomass of tropical forests: A new estimate based on forest volumes. *Science,* **223**, 1290-1293.

Brown, S., A.J.R. Gillespie and A.E. Lugo, 1989: Biomass estimation methods for tropical forests with applications to forest inventory data. *Forest Science,* **35**, 881-902.

Brown, S., 1991: Personal Communication.

Bulatao, R.A., E. Bos and M.T. Vu, 1989: *Asia Region Population Projections: 1980-80 Edition; Latin America and the Caribbean (LAC) Region Population Projections: 1980-80 Edition; Africa Region Population Projections: 1980-80 Edition; Europe, Middle East, and Africa (EMN) Region Population Projections: 1980-80 Edition.* Population and Human Resources Department, World Bank, Washington, D.C.

Burniaux, J.-M., J.P. Martin, G. Nicoletti and J.O. Martins, 1991: *The Costs of Policies to Reduce Global Emission of CO₂: Alternate Scenarios with GREEN.* Paris, September 1991 (Revised version II, 1991).

CEC (Commission of the European Communities), 1989: *Energy in Europe: Major Themes in Energy.* Brussels.

Crutzen, P.J. and M.O. Andreae, 1990: Biomass burning in the tropics: impact on atmospheric chemistry and biogeo-chemical cycles. *Science,* **250**, 1669-1678.

Edmonds, J. and J. Reilly, 1985: *Global Energy: Assessing the Future.* Oxford University Press, New York.

Edmonds, J.A., J.M. Reilly, R.H. Gardner and A. Brenkert, 1986: *Uncertainty in Future Global Energy Use and Fossil Fuel CO₂ Emissions 1975 to 2075.* TR036, DOE3/Nbb-0081 Dist. Category UC-11, National Technical Information Service, U.S. Department of Commerce, Springfield, Virginia.

Edmonds, J.A, I. Mintzer, W. Pepper, K. Major, V. Schater, M. Wise and R. Baron, 1992: *A Comparison of Reference Case Global Fossil Fuel Carbon Emissions.* Washington, D.C.

EPA (Environmental Protection Agency), 1990: *Policy Options for Stabilizing Global Climate: Report to Congress - Technical Appendices.* United States Environmental Protection Agency, Washington, D.C.

EPA (Environmental Protection Agency), 1991: *Global Inventory of Volatile Organic Compound Emissions From Anthropogenic Sources.* EPA, Air and Energy Engineering Research Laboratory, Research Triangle Park, North Carolina.

FAO (Food and Agriculture Organization), 1986: *1985 FAO Production Yearbook,* **Vol. 39**. FAO, Rome.

FAO (Food and Agriculture Organization), 1988: *An Interim Report on the State of the Forest Resources in the Developing Countries.* FAO, Rome.

FAO (Food and Agriculture Organization), 1991: *Forest Resources Assessment 1990 Project,* Forestry N. 7. FAO, Rome.

Grossling, B.F. and D.T. Nielson, 1985: *In Search of Oil, Volume 1: The search for oil and its impediments.* Financial Times Business Information, London.

Houghton, R.A., 1991: Tropical deforestation and atmospheric carbon dioxide. *Clim. Change* (Submitted).

IEA (International Energy Agency), 1991: *International Energy Agency: Energy and Oil Outlook to 2005.* IEA, Paris.

IPCC (Intergovernmental Panel on Climate Change), 1990a: *Climate Change: The IPCC Scientific Assessment.* J.T. Houghton, G.J. Jenkins and J.J. Ephraums (Eds.). WMO/UNEP. Cambridge University Press, Cambridge, UK. 365pp.

IPCC (Intergovernmental Panel on Climate Change), 1990b: *Emissions Scenarios prepared by the Response Strategies Working Group of the Intergovernmental Panel on Climate Change.* Report of the Expert Group on Emissions Scenarios.

IPCC (Intergovernmental Panel on Climate Change), 1991a: *Climate Change: The IPCC Response Strategies.* Island Press, Washington, D.C.

IPCC (Intergovernmental Panel on Climate Change), 1991b: *Energy and Industry Subgroup Report.* US Environmental Protection Agency, Washington, D.C.

IPCC (Intergovernmental Panel on Climate Change) Working Group I, 1991c: Summary of WGI task force meeting (Shepperton, UK, 8-11, July 1991).

Manne, A.S. and R.G. Richels, 1990: *Global CO₂ Emissions Reductions - the Impacts of Rising Energy Costs.* Electric Power Research Institute, Palo Alto, CA.

Masters, C.D., D.H. Root and E.D. Attanasi, 1991: Resource constraints in petroleum production potential. *Science*, **253**, 146-152.

Nordhaus, W.D. and G.W. Yohe, 1983: Future Carbon Dioxide Emissions from Fossil Fuels. In: *Changing Climate*. National Academy Press, Washington D.C., pp87-153.

OECD (Organization for Economic Cooperation and Development), 1991: *Estimation of Greenhouse Gas Emissions and Sinks*. Final Report from the OECD Experts Meeting, Paris, 18-21 February, 1991, revision of August 1991.

Pepper, W.J., J.A. Leggett, R.J. Swart, J. Wasson, J. Edmonds and I. Mintzer, 1992: *Emission Scenarios for the IPCC- an update:Background Documentation on Assumptions, Methodology, and Results*. US EPA, Washington, D.C.

Spiro, P.A., D.J. Jacob and J.A. Logan, 1991: Global inventory of sulphur emissions with 1°×1° resolution. *J. Geophys. Res.* (Submitted).

Swart, R.J., W.J. Pepper, C. Ebert and J. Wasson, 1991: *Emissions Scenarios for the Intergovernmental Panel on Climate Change - an Update: Background Paper and Workplan*. US EPA, Washington, D.C.

United Nations, 1990: *Population Prospects 1990*. United Nations, New York.

United Nations, 1992: *Long-Range World Population Projections*, United Nations Population Division, New York.

WEC (World Energy Conference), 1989: *Global Energy Perspectives 2000-2020*. WEC 14th Congress, Conservation and Studies Committee, Montreal.

Weyant, J., 1991: Stanford Energy Modeling Forum. Personal communication.

WMO, 1992: *Scientific Assessment of Ozone Depletion*. WMO/UNEP, WMO Global Ozone Research and Monitoring Project, Report No. 25, Geneva.

World Bank, 1991: *World Development Report 1991*. Oxford University Press, New York.

Zachariah, K.C. and M.T. Vu, 1988: *World Population Projections 1987-88 Edition*. Johns Hopkins University Press, Baltimore, Maryland.

B

Climate Modelling, Climate Prediction and Model Validation

W.L. GATES, J.F.B. MITCHELL, G.J. BOER, U. CUBASCH, V.P. MELESHKO

Contributors:
D. Anderson; W. Broecker; D. Cariolle; H. Cattle; R.D. Cess; F. Giorgi; M.I. Hoffert; B.G. Hunt; A. Kitoh; P. Lemke; H. Le Treut; R.S. Lindzen; S. Manabe; B.J. McAvaney; L. Mearns; G.A. Meehl; J.M. Murphy; T.N. Palmer; A.B. Pittock; K. Puri; D.A. Randall; D. Rind; P.R. Rowntree; M.E. Schlesinger; C.A. Senior; I.H. Simmonds; R. Stouffer; S. Tibaldi; T. Tokioka; G. Visconti; J.E. Walsh; W.-C. Wang; D. Webb

CONTENTS

EXECUTIVE SUMMARY

Coupled Model Experiments

- The new transient climate simulations with coupled atmosphere-ocean general circulation models (GCMs) generally confirm the findings of Section 6 in the 1990 Intergovernmental Panel on Climate Change report (IPCC, 1990), although the number of coupled model simulations is still small.

- There is broad agreement among the four current models in the simulated large-scale patterns of change and in their temporal evolution.

- The large-scale patterns of temperature change remain essentially fixed with time, and they become more evident with longer averaging intervals and as the simulations progress.

- The large-scale patterns of change are similar to those obtained in comparable equilibrium experiments except that the warming is retarded in the southern high latitude ocean and the northern North Atlantic Ocean where deep water is formed.

- All but one of the models show slow initial warming (which may be an artefact of the experimental design) followed by a nearly linear trend of approximately 0.3°C per decade.

- All models simulate a peak-to-trough natural variability of about 0.3°C in global surface air temperature on decadal time-scales.

- The rate of sea level rise due to thermal expansion increases with time to between 2 and 4cm/decade at the time of doubling of equivalent CO_2.

Regional Changes

- Although confidence in the regional changes simulated by GCMs remains low, progress in the simulation of regional climate is being obtained with both statistical and one-way nested model techniques. In both cases the quality of the large-scale flow provided by the GCM is critical.

Climate Feedbacks and Sensitivity

- There is no compelling new evidence to warrant changing the equilibrium sensitivity to doubled CO_2 from the range of 1.5 to 4.5°C as given by IPCC 1990.

- There is no compelling evidence that water vapour feedback is anything other than the positive feedback it has been conventionally thought to be, although there may be difficulties with the treatments of upper-level water vapour in current models.

- The effects of clouds remain a major area of uncertainty in the modelling of climate change. While the treatment of clouds in GCMs is becoming more complex, a clear understanding of the consequences of different cloud parametrizations has not yet emerged.

Atmospheric Variability

- Model experiments with doubled CO_2 give no clear indication of a systematic change in the variability of temperature on daily to interannual time-scales, while changes of variability for other climate features appear to be regionally (and possibly model) dependent.

Ocean Modelling

- Results from eddy-resolving ocean models show a broadly realistic portrayal of oceanic variability, although the climatic role of eddies remains unclear.

- Ocean-only GCMs show considerable sensitivity of the thermohaline circulation on decadal and longer time-scales to changes in the surface fresh-water flux, although coupled models may be less sensitive.

- Sea-ice dynamics may play an important role in the freezing process, and should therefore be included in models for the simulation of climate change under increased CO_2.

Model Validation

- There have been improvements in the accuracy of individual atmospheric and oceanic GCMs, although the ranges of intermodel error and sensitivity remain large.

- The lack of adequate observational data remains a serious impediment to climate model improvement.

B1 Introduction

Since the publication of the first IPCC Scientific Assessment of Climate Change (IPCC, 1990) there have been significant advances in many areas of climate research as part of a continuing worldwide acceleration of interest in the assessment of possible anthropogenic climate changes. In this section we concentrate on advances in the modelling of climate change due to increased greenhouse gases, improvements in the analysis of climate processes and feedbacks, and on advances in climate model validation that were not available to the IPCC in early 1990. This section is thus intended as an update to selected portions of the 1990 assessment rather than as a comprehensive revision, and an effort has been made to keep the discussion both concise and focussed in accordance with the stringent space limitations placed upon this supplementary report.

In Section B2 recent simulations of climate change are assessed, with emphasis on results from coupled ocean-atmosphere models. In Section B3, recent research on modelling climate feedbacks is discussed, and the IPCC 1990 estimates of climate sensitivity are reviewed. The simulation of atmospheric variability and its changes due to increased atmospheric CO_2 are discussed in Section B4 while developments in ocean and sea-ice modelling are presented in Section B5. Finally, Section B6 discusses advances in climate model validation.

B2 Advances in Modelling Climate Change due to Increased Greenhouse Gases

B2.1 Introduction

Simulation of the climatic response to increases in atmospheric greenhouse gases has continued to dominate climate modelling. Preliminary results from new integrations of coupled global atmosphere-ocean models with progressive increases of CO_2 show that the patterns of the transient response are similar to those in an equilibrium response, except over the high-latitude southern ocean and northern North Atlantic ocean; here the delayed warming has highlighted the critical role of the oceanic thermohaline circulation. Computing limitations have continued to restrict the resolution that can be used in GCM simulations of the climate changes due to increased greenhouse gases, although progress is being made in the simulation of regional climate by both statistical techniques and by locally nesting a higher-resolution model within a global GCM.

B2.2 New Transient Results from Coupled Atmosphere-Ocean GCMs

At the time of the 1990 IPCC report, preliminary results were available from only two coupled model integrations with transient CO_2, namely those made at NCAR (Washington and Meehl, 1989) and at GFDL (Stouffer *et al.*, 1989). Of these only the GFDL integration had been carried to the point of CO_2 doubling (which occurred after

Table B1: *Summary of transient CO_2 experiments with coupled ocean-atmosphere GCMs*

	GFDL	MPI	NCAR	UKMO
AGCM	R15 L9	T21 L19	R15 L9	2.5° × 3.75° L11
OGCM	4.5° × 3.75° L12	4° L11	5° L4	2.5° × 3.75° L17
Features	no diurnal cycle, isopycnal ocean diffusion	prognostic CLW, quasi-geostrophic ocean		prognostic CLW, isopycnal ocean diffusion
Flux adjustment	seasonal, heat, fresh water	seasonal, heat, fresh water, wind stress	none	seasonal, heat, fresh water
Control CO_2 (ppm)	300	390 †	330	323
CO_2 (t)	1% yr⁻¹ (compound)	IPCCa & d (1990 Scenarios), 2×CO_2	1% yr⁻¹ (linear)	1% yr⁻¹ (compound)
Length (yr)	100	100	60	75
CO_2 doubling time (yr)	70	60 (IPCCa)	100	70
Warming (°C) at CO_2 doubling	2.3	1.3	2.3 (projected)	1.7
2×CO_2 sensitivity (°C) (with mixed layer ocean)	4.0	2.6	4.5	2.7 ††

L - number of vertical levels; R - number of spectral waves (rhomboidal truncation); T - no spectral waves (triangular truncation).

  † - equivalent CO_2 value for trace gases
  †† - estimate from low resolution experiment

approximately 70 years as a result of a 1% per year compound increase of CO_2). Here we present further results from the transient CO_2 experiments with both the GFDL and NCAR coupled models, along with preliminary results from new transient CO_2 integrations recently completed at the Max-Planck-Institute for Meteorology (MPI) in Hamburg and at the Hadley Centre of the United Kingdom Meteorological Office (UKMO). A fuller description and intercomparison of these results is being prepared (WCRP, 1992).

A summary of the new transient results is presented in Table B1. Note that there are differences in the CO_2 equivalent doubling times (by a factor of almost two) and that the equilibrium sensitivity of the models is different. The UKMO model has the finest horizontal resolution while the MPI model has the finest vertical resolution in the atmosphere and includes explicitly the radiative effects of other trace gases. Three of the four coupled models (see Table B1) use a correction or adjustment of the air-sea fluxes so that the CO_2-induced changes should be interpreted as perturbations around a climate similar to that presently observed. Adjustment of the fluxes also prevents distortion of the CO_2 induced perturbation by rapid drift of the model climate (since the same corrections are applied to the control and anomaly simulations).

1. The new transient climate simulations with coupled atmosphere-ocean GCMs generally confirm the findings of Section 6 in the 1990 IPCC report, although the number of coupled model simulations is still small.

The globally averaged annual mean increase of surface air temperature at the time of effective CO_2 doubling (70 years for the GFDL and UKMO models, 60 years for the MPI model with IPCC Scenario A, and 100 years for the NCAR model) is between 1.3°C and 2.3°C. These values are approximately 60% of the models' equilibrium warming (where known) with doubled CO_2 when run with simple mixed-layer oceans. These results confirm those from the simple models used in IPCC 1990. The lower values of warming compared to the equilibrium experiments are partly due to the fact that the transient experiments take into account the thermal inertia of the deep ocean and are therefore not at equilibrium at the time of effective CO_2 doubling.

2. All but one of the models show a limited initial warming (which may be an artefact of the experimental design) followed by a nearly linear trend of approximately 0.3°C per decade. All models simulate a peak-to-trough natural variability of about 0.3 to 0.4°C in global surface air temperature on decadal time-scales.

The evolution of the change of globally averaged surface air temperature (sea surface temperature for the NCAR experiment) during the course of the various transient experiments with coupled ocean-atmosphere

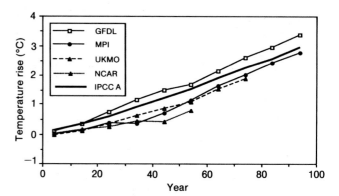

Figure B1: Decadal mean changes in globally averaged surface temperature (°C) in various coupled ocean-atmosphere experiments. (see Table B1). Note that the scenarios employed differ from model to model, and that the effect of temperature drift in the control simulation has been removed. Open boxes = GFDL; solid circles = MPI; triangles with dashed line = UKMO; triangles with dotted line = NCAR (sea temperatures only); solid line = IPCC 1990 Scenario A "best estimate".

GCMs is shown in Figure B1. In spite of the differences in the models' parametrizations and in their experimental configurations, all of the models exhibit a number of similar overall features in their response. Firstly, three of the models show relatively little warming during the first few decades of the integration rather than a constant rate of warming throughout, despite the near constant rate of increase in radiative heating. This so-called "cold start " is barely noticeable in the GFDL simulation, but in the UKMO and MPI models the warming is negligible during the first 2 to 3 decades. This phenomenon is thought to be an artefact of the experimental design and can be reproduced qualitatively using simplified models (Hasselmann *et al.*, 1992; J.M. Murphy, personal communication; see also Hansen *et al.*, 1985) which indicate that the length of delay grows with increasing model sensitivity (the equilibrium warming for doubling CO_2) and with effective heat capacity in the ocean. Until this "cold start" phenomenon is investigated and understood, it is not meaningful to match "model time" with calendar dates. Secondly, as the increase in CO_2 progresses, each model approximates a constant rate of warming, in overall agreement with the 0.3°C/decade in the IPCC (1990) projections made for the "business-as-usual" CO_2 forcing scenario (A) with a simplified upwelling-diffusion ocean model. Thirdly, all models exhibit variability on interannual, decadal and even longer time-scales, and some models reproduce ENSO-like effects (Meehl,1990; Lau *et al.*, 1991; J.F.B. Mitchell, personal communication). The peak-to-trough range of global surface air temperature intra-decadal natural variability is 0.3 to 0.4°C (see, for example, Figure B2).

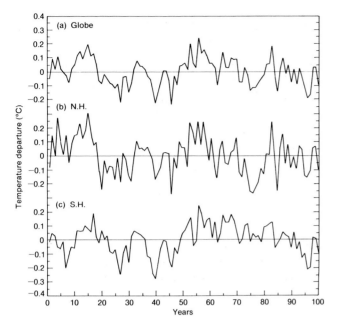

Figure B2: Temporal variations of the area-averaged deviation of annual mean surface air temperature (°C) from the corresponding 100-year average of the control as simulated by the GFDL coupled ocean-atmosphere model for: (a) the globe, (b) the Northern Hemisphere, and (c) the Southern Hemisphere. (From Manabe *et al.*, 1991.)

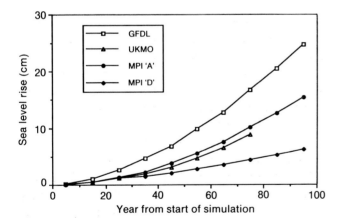

Figure B3: Decadal mean changes in globally averaged sea level change (cm) from various coupled ocean-atmosphere GCM experiments (see Table B1). Open squares = GFDL; solid circles = MPI (IPCC Scenario "A"); solid diamonds = MPI (IPCC Scenario "D"); open triangles = UKMO. Note the differences in forcing in Table B1, and that the effect of long-term drift in the models has been removed by differencing.

Analysis of the nature of this variability in the coupled integrations and of the implications for detecting global warming is currently in progress.

3. The rate of sea level rise due to thermal expansion increases with time to between 2 and 4cm/decade at the time of doubling of equivalent CO_2.

The "cold start" leads to a substantial delay in the associated change in global sea level rise due to thermal expansion (Figure B3), and is most pronounced in the UKMO and MPI models. The rate of sea level rise increases during the simulations to 2, 2.5 and 4cm/decade by the time of CO_2 doubling in the MPI A (Scenario A), UKMO and GFDL experiments, respectively. The simulated rate of sea level rise at 2030 (the time of doubling of CO_2) due to thermal expansion (Scenario "A" best estimate) in the previous IPCC assessment was about 2.5cm/decade. Note that we have not reassessed the contribution to sea level rise from changes in the snow and ice budgets over land.

4. The large-scale patterns of change are similar to those obtained in comparable equilibrium experiments except that the warming is retarded in the southern high latitude ocean and the northern North Atlantic Ocean where deep water is formed.

The geographical distribution of the change in surface air temperature at the approximate time of CO_2 doubling in the four transient experiments is shown in Figure B4. (For the NCAR model the CO_2 increase is only 60%) The same overall characteristics are seen in the distribution of the simulated change of surface air temperature. These characteristics are: (1) the largest warming occurs in the high latitudes of the Northern Hemisphere, (2) relatively uniform warming occurs over the tropical oceans, and (3) a minimum of warming or in some cases cooling occurs over the northern North Atlantic and over the Southern Ocean around Antarctica. Features (1) and (2) are familiar from the earlier studies of the equilibrium warming in response to doubled CO_2 with a mixed-layer ocean (see Section B2.3); the high latitude ocean warming minima, on the other hand, are due to the presence of upwelling and deep vertical mixing which increase the effective heat capacity and hence the thermal inertia of the ocean locally.

5. There is broad agreement among the four models in the simulated large-scale patterns of change and in their temporal evolution. The large-scale patterns of temperature change remain essentially fixed with time, and they become more evident with longer averaging intervals and as the simulations progress.

A larger warming over land areas compared to that over the oceans is common to all models in both winter and summer. This ocean-land asymmetry contributes to a more rapid warming in the Northern Hemisphere compared to that in the Southern Hemisphere. There is a maximum warming over the Arctic Ocean in winter (Figure B5a) and a minimum in summer (Figure B5b), similar to the equilibrium results. However, the warming and its seasonal

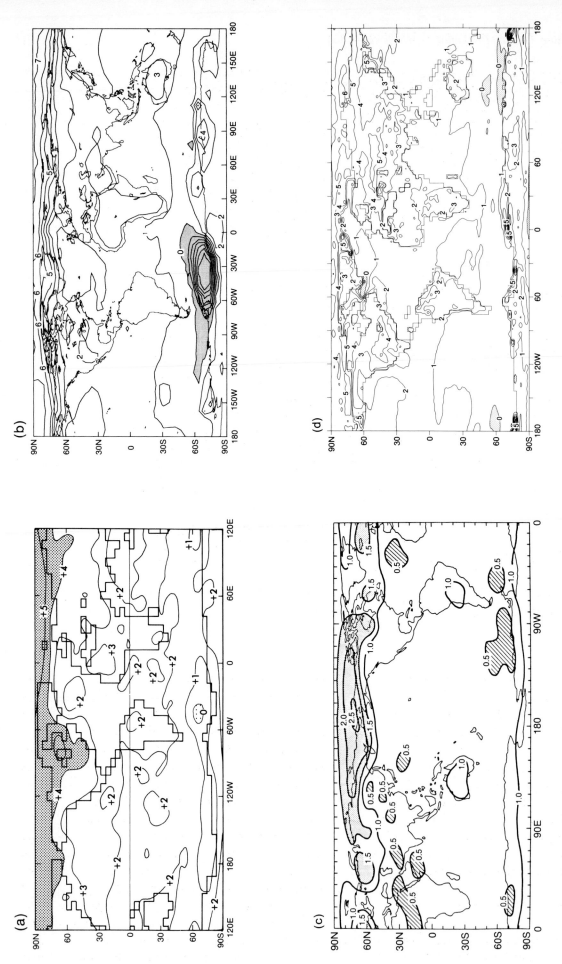

Figure B4: The distribution of the change of surface air temperature (°C) simulated near the time of CO_2 doubling by four coupled ocean-atmosphere GCMs in response to a transient CO_2 increase. (a) The GFDL results are averaged over years 60-80 and referenced to the 100-year average of a control; (b) the MPI results are averaged over years 56-65 and referenced to the corresponding years of a control; (c) the NCAR results are averaged over years 31-60 and referenced to the corresponding control years; (d) the UKMO results are averaged over years 65-75 and are referenced to the corresponding years of a control. (Manabe *et al.*, 1991; U. Cubasch, G.A. Meehl and J.F.B. Mitchell, all by personal communication.)

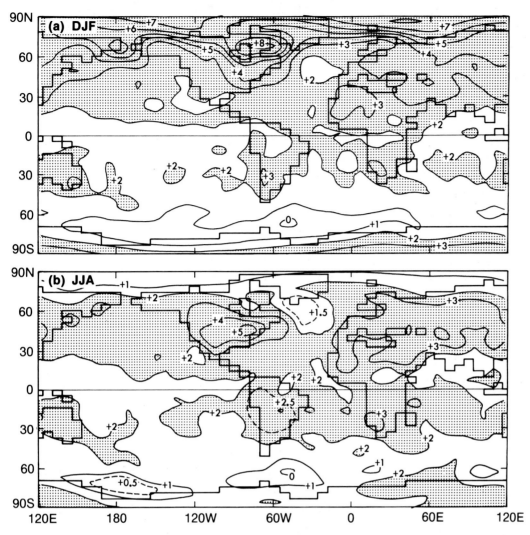

Figure B5: Distribution of the mean surface air temperature (°C) for (a) DJF, and (b) JJA during years 60-80 of a transient CO_2 simulation with the GFDL model, relative to the 100 year average of the control. (From Manabe *et al.*, 1992.)

variation over the circumpolar ocean in the Southern Hemisphere is considerably less than in the equilibrium models because of deep vertical mixing in the ocean. Precipitation increases in the Northern Hemisphere in high latitudes throughout the year, in much of mid-latitudes in winter, and in the southwest Asian monsoon (Figure B6a, b). In the Southern Hemisphere, there are precipitation increases along the mid-latitude storm tracks throughout the year. In the Northern Hemisphere, winter soil moisture increases over the mid-latitude continents, while in the summer there are many areas of drying; this is also similar to the results obtained by equilibrium models (Figure B7a, b).

In at least one simulation (Cubasch *et al.*, 1991) the progressive change in surface temperature is unrelated to the pattern of internal variability in the model; the first empirical orthogonal function (EOF) of the annual mean surface temperature in the control run is uncorrelated with the first EOF from the simulation with increasing

greenhouse gas concentrations. The climate change pattern becomes more evident as the integration progresses, as indicated by the growth in the fraction of variance explained by the first EOF of the anomaly experiment (Figure B8). This feature is noticeable in other models (for example, Meehl *et al.*, 1991a) and in other fields such as soil moisture (Figure B9) and (to a lesser extent) in the precipitation (Santer *et al.*, 1991; J.F.B. Mitchell, personal communication). In IPCC 1990 it was assumed that the regional climate changes would scale linearly with the global mean temperature changes; while Figure B9 offers some support for this assumption, there is considerable interdecadal variability at regional scales in the coupled model simulations, so this scaling remains questionable.

There are a number of caveats to be kept in mind in assessing the results of the transient CO_2 simulations with coupled ocean-atmosphere models shown here. Chief among these is the use (in the GFDL, MPI and UKMO models) of adjustments to the oceanic surface fluxes so

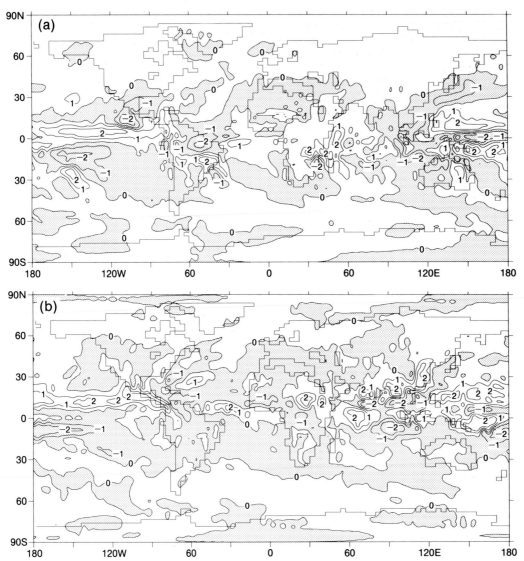

Figure B6: Decadally averaged changes in precipitation (mm/day) for (a) DJF, and (b) JJA around the time of doubling of CO_2 in an experiment in which CO_2 was increased by 1%/year in the UKMO model (J.F.B. Mitchell, personal communication). Contours are every 1mm/day and areas of decrease are stippled.

that the ocean temperature and salinity remain close to present climatology. If flux corrections are not used (as in the NCAR model), imperfections in the component models (and in their interaction) may introduce significant systematic errors in the coupled simulation. Such flux corrections or adjustments represent substantial changes of the fluxes that are exchanged between the component models and, as shown in Figure B10, are comparable in magnitude to the atmospheric fluxes. In the case shown, they tend to be of opposite sign, i.e., the flux correction substantially reduces the net effective flux to the ocean. The use of such adjustments may distort the models' response to small perturbations like those associated with increasing CO_2. On the other hand, the similarities in the responses of the NCAR model without flux corrections and in the three models that use flux corrections indicate that the simulated changes in the models may not be

substantially affected by flux adjustment. In addition, confidence in the overall validity of the results is raised by the consistency of parallel experiments with the GFDL model in which the CO_2 undergoes a progressive transient reduction (Manabe *et al.*, 1991,1992).

B2.3 New Equilibrium Results from Atmospheric GCMs and Mixed-Layer Ocean Models

Although increasing attention is being paid to transient simulations with coupled atmosphere-ocean GCMs (see Section B2.2), doubled CO_2 experiments with atmospheric GCMs and mixed-layer ocean models continue to be of interest since they can be carried to statistical equilibrium relatively easily and they provide a benchmark for model sensitivity. The results of several new such experiments completed since the publication of the 1990 IPCC Scientific Assessment (IPCC, 1990) are summarized in

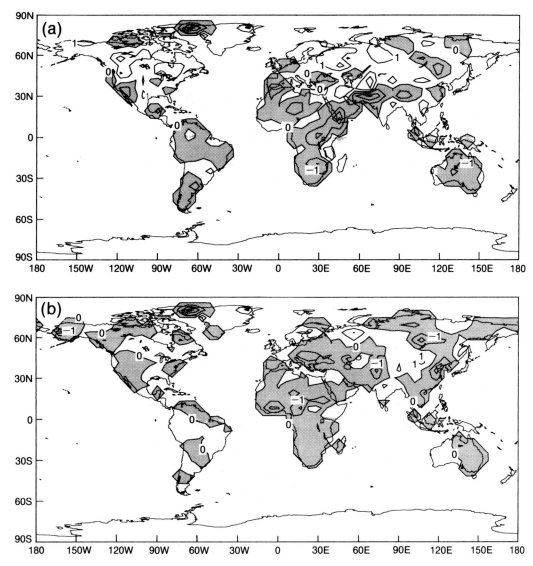

Figure B7: Decadally averaged changes in soil moisture in the MPI model for (a) DJF, and (b) JJA around the time of doubling of effective CO_2. Contour intervals are every cm. Areas of decrease are shaded. (Cubasch *et al.*, 1991.)

Table B2. In general the simulated globally-averaged increases in surface air temperature and precipitation are similar to those found earlier (see IPCC, 1990, Table 2.3 (a); the differences relative to earlier models can be related in some cases to the differences in cloud radiative feedback in the tropics (see Section B3.3)). The largest changes are found in the one new model (LMD) that includes prognostic cloud liquid water and variable cloud optical properties.

New model results suggest that the difference in the vertical distribution of the radiative forcing has implications for the simulations of the control climate as well as for the greenhouse gas-induced warmer climates. As demonstrated by Wang *et al.* (1991b), inclusion of individual trace gases (i.e., not as equivalent CO_2) in a control simulation produces a warmer atmosphere, especially in the tropical upper troposphere. Since this model (and most other GCMs without trace gases) show a

cold bias relative to observations of the current climate (Boer *et al.*, 1991a; see Section B5.2), inclusion of the trace gases should tend to reduce this systematic error. Note, however, that the anticipated cooling effect of aerosols (Charlson *et al.*, 1991) is not included in current climate models.

The regional climate responses due to the inclusion of additional greenhouse gases have been examined by Wang *et al.* (1992). In one experiment the trace gases (CO_2, CFCs, methane and nitrous oxide) were represented explicitly, while in the second trace gases were represented by adding the amount of CO_2 which gives an equivalent increase in radiative heating at the tropopause, i.e., using CO_2 as a surrogate for the trace gases. When the trace gases and CO_2 are given values in 2050 according to the IPCC "business-as-usual" scenario (at which time the concentrations of CO_2, CH_4, N_2O, CFC-11 and CFC-12 would have increased by 52, 89, 19, 49 and 121%,

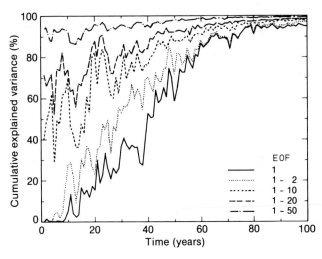

Figure B8: Cumulative spatial variance explained as a function of time and number of EOFs (1, 2, 10, 20 and 50) for Scenario "A" (see text for explanation). The signal is defined as the difference relative to the smoothed initial state (average over years 1-10) of the control run of the MPI model. Results are for annually averaged 2m temperature. (From Cubasch *et al.*, 1991.)

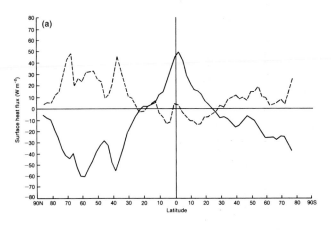

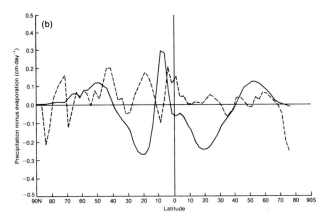

Figure B10: The zonal mean of the surface fluxes of (a) heat, and (b) precipitation minus evaporation, simulated by the atmosphere in the UKMO coupled GCM (solid line) and the flux corrections (dashed line) added during the course of the coupled model integration. (From Murphy, 1990.)

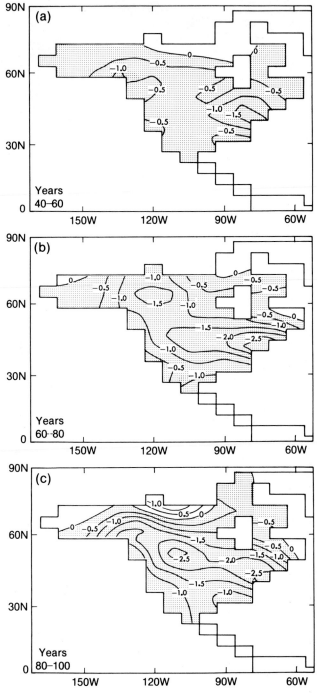

Figure B9: Distribution of the change in JJA soil moisture (cm) simulated during successive 20-year periods over North America relative to a control by the GFDL coupled model with a 1% per year (compound) increase of CO_2: (a) years 40 to 60; (b) years 60 to 80, and (c) years 80 to 100. (From Manabe *et al.*, 1992.)

Table B2: *Summary of results from new equilibrium simulations for doubled CO_2 with atmospheric GCMs with a seasonal cycle and a mixed-layer ocean.*

Group	Investigators	Year	Resolution	Diurnal cycle	Convection	Cloud scheme	Cloud optical property	ML depth (m)	Ref (CO_2) (ppm)	ΔT (°C)	ΔP (%)	Simulation length (years)
BMRC	McAvaney et al.	1991	R21 L9	yes	PC	RH	fixed	50	330	2.2	3.0	15
YALE/ CCM1	Oglesby & Saltzman	1990	R15 L12	no	MCA	RH	fixed	50	330	3.8	n.a.	100
SUNY A/ CCM1	Wang et al.	1991b	R15 L12	no	MCA	RH	fixed	50	330	4.2	8.3	20
CSIRO	Dix et al.	1991	R21 L9	yes	PC	RH	fixed	50	330	4.8	10	30
NCAR/ CCM	Washington & Meehl	1991	R21 L9	no	MCA	RH	fixed	50	330	4.5	5	50
SUNY A/ CCM1	Wang et al.	1992	R15 L12	no	MCA	RH	fixed	50	354 [†]	4.0 [††] 3.9 [†††]	7.1 [††] 6.9 [†††]	100 20
LMD	Le Treut et al.	1992	5°x7.5° L11	no	MCA & KUO	CW	variable	50	320	5.3	8	20
IAP	Wang et al.	1991a	4°x5° L2	yes	MCA	RH	fixed	60	324	1.7	2.5	16

PC - Penetrative convection;   MCA - Moist convective adjustment;   RH - Relative humidity;   CW - Cloud water;   ML - Mixed layer.
L - number of vertical levels;   R - number of spectral waves (rhomboidal truncation).

[†]   -   1990 CO_2 concentrations in SA90, which also includes the explicit trace gas concentrations (CH_4) = 1.717 ppm, (N_2O) = 0.309 ppm, ($CFCl_3$) = 0.280 ppm and (CF_2Cl_2) = 0.484 ppb.

[††] -   Changes between equilibrium solutions at 2050 relative to 1990, for which the concentration of CO_2 increased by 52%, CH_4 by 89%, N_2O by 19%, $CFCl_3$ by 49% and CF_2Cl_2 by 121% according to SA90.

[†††] -   As in [††] except for equivalent CO_2. The concentration of equivalent CO_2, 660 ppm, is calculated from equating the increased global mean longwave radiative forcing for the troposphere–surface system.

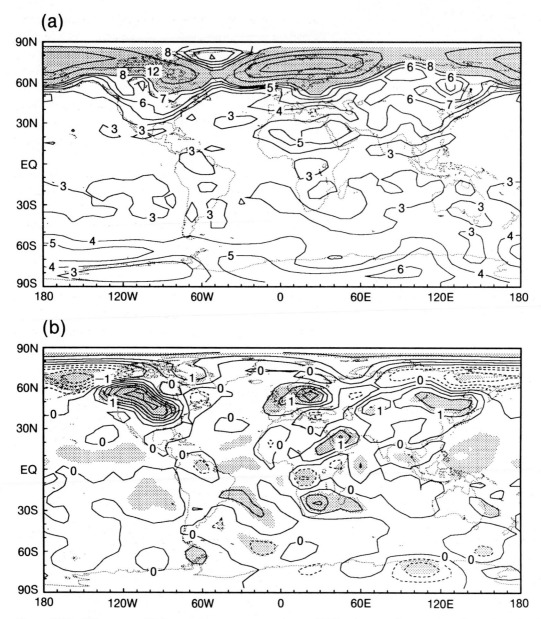

Figure B11: Effect on equilibrium surface temperature change (DJF) using actual trace gases (Scenario "A", 1990-2050) instead of radiatively equivalent increases in CO_2 (Wang *et al.*, 1992): (a) warming using actual trace gases. Contour interval is 1°C except in stippled region (>8°C) where the interval is 4°C; (b) difference due to use of actual trace gases rather than effective CO_2. Contour interval is 0.5°C, negative contours are dashed and areas of statistical significance are stippled.

respectively, relative to 1990), the equilibrium simulations show nearly identical annual and global mean surface warming and increased precipitation (see Table B2). However, the regional pattern of climate changes between the two experiments is different (Figure B11). Although the physical mechanisms for these differences have not yet been established, such results suggest that atmospheric trace gases other than CO_2 should be included explicitly in future GCM simulations of both the present and future climates.

B2.4 Regional Climate Simulations
Although confidence in the regional changes simulated by

GCMs remains low, progress in the simulation of regional climate is being made with both statistical and one-way nested model techniques. In both cases the quality of the large-scale flow provided by the GCM is critical.

The horizontal resolution in current general circulation models is too coarse to provide the regional scale information required by many users of climate change simulations. The development of useful estimates of regional scale changes is dependent upon the reliability of GCM simulations on the large-scale. Recognizing that simple interpolation of coarse-grid GCM data to a finer grid is inadequate (Grotch and MacCracken, 1991), strategies have been developed that involve the projection

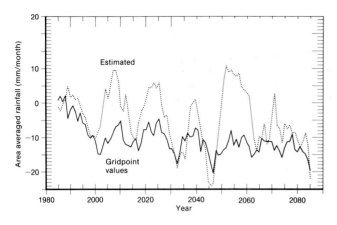

Figure B12: Projected 5-year mean changes in precipitation over the Iberian peninsula (mm/month) in a transient CO_2 experiment. Solid line = grid point values from a GCM; dashed lines = values from a statistical approach with large-scale patterns derived from the GCM. (From Von Storch *et al.*, 1992.)

of large-scale information from GCMs onto the regional scale either by using limited area models with boundary conditions obtained directly from the GCM (the one-way nested approach) or by using empirically-derived relationships between regional climate and the large-scale flow (the statistical approach). In both approaches, however, accuracy is limited by the accuracy of the large-scale flow generated in the GCM, and so they do not avoid the need for improvement in GCM simulations.

Since the preparation of the IPCC (1990) report there has been considerable research using statistical approaches to climate simulation; see, for example, Wigley *et al.* (1990), Karl *et al.* (1990) and Robock *et al.* (1991) amongst others. The statistical technique of Von Storch *et al.* (1991) has successfully reproduced observed rainfall patterns over the Iberian peninsula, and has been applied to an IPCC Scenario A model integration (Cubasch *et al.*, 1991). The resulting changes in the winter rainfall are shown in Figure B12 as compared to the original GCM

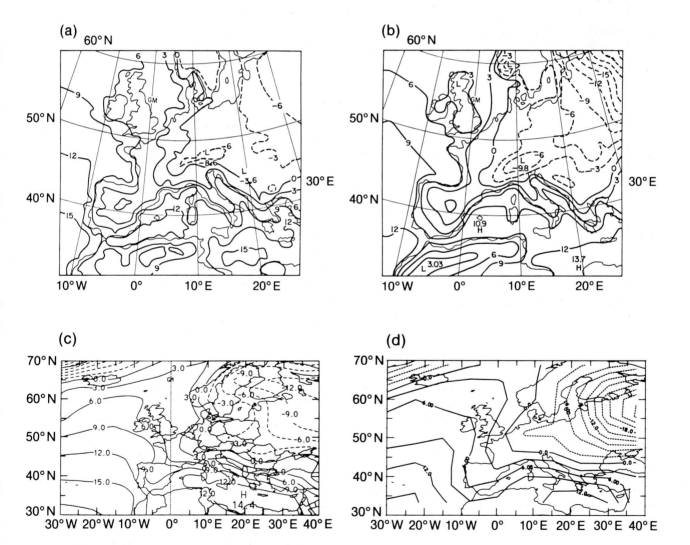

Figure B13: The average January surface air temperature (°C) over Europe: (a) as given by high-resolution observations; (b) as simulated by a mesoscale model nested within the NCAR CCM1 R15; (c) as given by large-scale observations, and (d) as simulated by the NCAR CCM1 alone. (From Giorgi *et al.*, 1990.)

grid-point rainfall amounts; here, however, only about half of the year-to-year variability is explained. Similar research is underway at other institutions (UKMO, BMRC). A recent review of research in these areas has been prepared by Giorgi and Mearns (1991).

The one-way nested modelling approach has been applied by Giorgi *et al.* (1990), whose regional scale climate simulations shown in Figure B13 compare well with high-resolution observations, and show a significant amount of detailed information that was not portrayed in the original GCM. (The limited area model is likely to produce significant improvements only in regions where topographic effects are substantial.) Similar results have been found by McGregor and Walsh (1991) with the CSIRO model and by other groups (MPI, UKMO). The application of the nesting approach to regional climate changes in increased CO_2 experiments is not quite so advanced, although Giorgi *et al.* (1991) have recently used the NCAR doubled CO_2 simulation of Washington and Meehl (1991) in this way; their simulated changes of January surface air temperature for both the original GCM and the GCM plus a one-way nested mesoscale model are shown in Figure B14.

In spite of this progress, the relatively low confidence attached to regional projections of any sort should be emphasized. Analysis by Karl *et al.* (1991) has shown that the observed changes of temperature and the winter-to-summer precipitation ratio over central North America are not consistent with the IPCC (1990) projections, though this inconsistency could be due to factors other than model errors.

B3 Advances in the Analysis of Climate Feedbacks and Sensitivity

B3.1 Introduction

While climate simulation continues apace, new attention has been given to the critical role of feedback processes in determining the climate's response to perturbations. In general, water vapour is expected to amplify global warming, while the effect of clouds remains uncertain. Overall, there is no compelling evidence to change earlier model estimates of the climate's sensitivity to increased greenhouse gases.

B3.2 Water Vapour Feedback

There is no compelling evidence that water vapour feedback is anything other than the positive feedback it has generally been considered to be, although there may be difficulties with the treatments of upper-level water vapour in current models.

The importance of water vapour feedback in climate change has long been recognized (Manabe and Wetherald, 1967) and the dependence of the current greenhouse effect on water vapour has been documented by Raval and

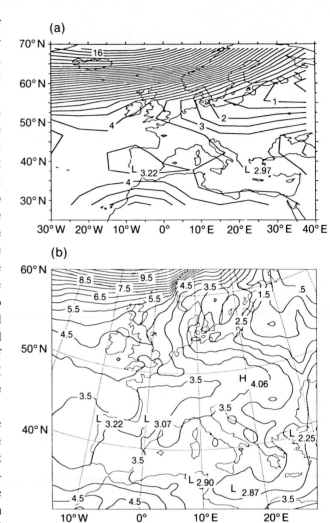

Figure B14: The distribution of changes in January surface air temperature over Europe as a result of doubled CO_2: (a) the temperature change (°C) given by the NCAR GCM relative to the GCM control; (b) the temperature change (°C) given by the nested mesoscale model relative to the nested model control. (From Giorgi *et al.*, 1991.)

Ramanathan (1989) using satellite data from regions of clear skies. The theoretical maximum concentration of water vapour is governed by the Clausius-Clapeyron relation, and increases rapidly with temperature (about 6%/°C); this is the physical basis for the strong positive water vapour feedback seen in present climate models (whereby increases in temperature produce increases in atmospheric water vapour which in turn enhance the greenhouse effect leading to a warmer climate). Increases in water vapour will be more effective at higher levels where temperatures are lower, and the change in emissivity (or effectiveness in absorbing thermal radiation) will be approximately proportional to the **fractional** change in water vapour. (This simple relationship is complicated by absorption by the water vapour continuum in the tropics, so that for a given percentage increase in water vapour, the

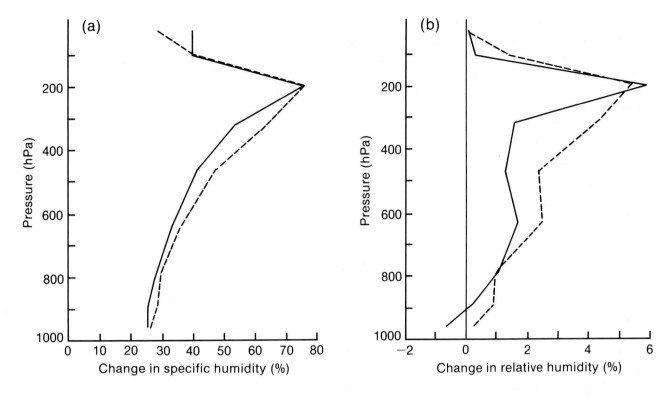

Figure B15: Vertical profiles of the global mean change in (a) specific humidity, and (b) relative humidity, simulated by the GISS GCM in a warm climate (+2°C SST change) relative to a cold climate (-2°C SST change). The full line denotes the standard version of the model and the dashed line denotes an improved version. (From Del Genio *et al.*, 1991.)

biggest contribution to increasing the greenhouse effect occurs from about 500 hPa in cold regions, but from about 800 hPa in the tropics (Shine and Sinha, 1991).)

The main source of water vapour is evaporation over the oceans, and humidities in the oceanic boundary layer are close to saturation, with the concentration decreasing rapidly with height. Both convective and large-scale motions and the evaporation of precipitation help to maintain the moisture distribution of the middle troposphere (Betts, 1990; Pierrehumbert, 1991; Kelly *et al.*, 1991). The generally strong vertical gradients of water vapour may, however, lead to systematic errors in models with parametrized vertical moisture diffusion or coarse vertical resolution. Consequently, the large-scale transport of moisture must be treated carefully in GCMs lest spurious sources and sinks of moisture occur (Rasch and Williamson, 1990, 1991; Kiehl and Williamson, 1991). Nevertheless, the overall realism of the outgoing clear-sky radiative fluxes and of the simulated mean tropospheric temperatures suggests that the models' simulated tropospheric water vapour concentrations are not greatly in error (Cess *et al.*, 1990; Boer *et al.*, 1991a).

Under global warming, there is general agreement that the atmospheric boundary layer would become moister, and as a result of both large-scale and convective mixing, the humidity in the upper troposphere is also likely to increase. All current GCMs simulate a strong positive

water vapour feedback (Cess *et al.*, 1990) with *relative* humidity remaining constant to a first approximation (Mitchell and Ingram, 1992). Lindzen (1990) noted, however, that at warmer temperatures, tropical convective clouds might detrain moisture at higher, colder temperatures, and thus provide less water to the atmosphere directly. Del Genio *et al.* (1991) diagnosed results from two versions of the GISS GCM, and found that in a warmer climate, cumulus-induced subsidence indeed tends to increase drying as anticipated by Lindzen, but found that this effect is offset by increased moistening as a result of upward moisture transport by large-scale eddies and by the tropical mean meridional circulation (Figure B15).

Estimates of the strength of the water vapour feedback can be made from observations, although they are not independent of the effects of atmospheric motions. As noted earlier, Raval and Ramanathan (1989) computed the dependence of the normalized clear-sky greenhouse effect derived from ERBE satellite data against sea surface temperature from different locations and seasons. The enhancement due to ascent in the intertropical convergence zone (at temperatures around 303K) and the reduction due to subsidence (at temperatures around 295K) can be seen in Figure B16. The overall positive slope suggests that the greenhouse effect indeed increases with temperature, although work by Arking (1991) indicates that some of the

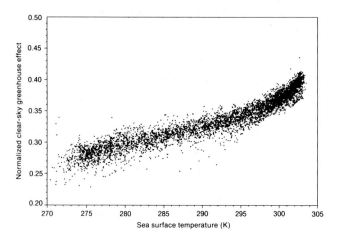

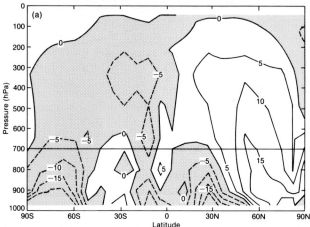

Figure B16: The dependence of the global clear-sky greenhouse effect normalized by infrared emission (g) as a function of sea surface temperature (T, in K) as given by ERBE data during April 1985.

contribution to the slope is due to changes in the vertical profile of temperature. Results from several GCMs have been found to agree with Raval and Ramanathan's results and with other satellite data. For example, Rind *et al.* (1991) compare the simulated July to January relative humidity difference simulated by the GISS model in cloud-free areas with the corresponding observations from the SAGE II instrument (which retrieves water vapour concentration between the stratopause and cloud top altitude). As shown in Figure B17, in the region above 700 hPa the model is in rather good agreement with the observations, especially in the tropical areas that are dominated (in the clear-sky areas) by subsidence from penetrating cumulus convection. (Although the simulated relative humidity was too high by about 30%, this need not necessarily exaggerate the water vapour feedback, as the feedback is approximately proportional to the fractional change in water vapour content.) The model also simulates the increased relative humidity in the middle and upper troposphere in summer and that in the convectively-dominated tropical western Pacific relative to the largely non-convective eastern Pacific.

B3.3 Cloud Feedback

The effects of clouds remain a major area of uncertainty in the modelling of climate change. While the treatment of clouds in GCMs is becoming more complex, a clear understanding of the consequences of different cloud parametrizations has not yet emerged.

Cloud feedback is the term used to encompass effects of changes in cloud and their associated radiation on a change of climate, and has been identified as a major source of uncertainty in climate models (see IPCC (1990) Sections 3 and 11). This feedback mechanism incorporates both

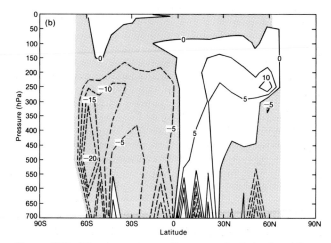

Figure B17: The change in zonally-averaged relative humidity (%) between July and January: (a) as simulated by the GISS GCM, and (b) as observed over a 5-year period by the satellite-borne SAGE II instrument. (From Rind *et al.*, 1991.)

changes in cloud distribution (horizontal and vertical) and changes in cloud radiative properties, although these cannot always be separated.

Observational data on the variation of cloud radiative properties in terms of other variables, such as cloud water content and temperature are relatively scarce. The observations of Feigelson (1978) over the Soviet Union, used by Somerville and Remer (1984), suggest a general increase of cloud optical thickness for temperatures below about 0°C, for which Betts and Harshvardhan (1987) provided plausible supporting thermodynamic arguments. The implication is that low clouds become more reflective as temperature increases, thereby introducing a negative feedback, while the feedback from high clouds depends upon their height and coverage and could be of either sign. The feedback effect of changes in temperature on low-cloud optical properties has recently been considered by Tselioudis *et al.* (1991) in a study based on ISCCP data. They infer a decrease in low-cloud optical depth with increasing temperature for warm continental and almost all maritime clouds. If this correlation holds for global warming, then it represents a positive feedback mechanism associated with low cloudiness in these regions. (Note, however, that other processes can lead to local changes in cloud optical depth, and the correlation with temperature may be fortuitous.) Feigelson's data, however, suggest little change in optical depth with increasing temperature for water clouds.

Ramanathan and Collins (1991) have discussed a negative feedback mechanism associated with the production of highly reflective cirrus clouds as a consequence of an increase in tropical convective activity associated with the 1987 El Niño. The importance of this mechanism for the case of greenhouse warming remains to be fully investigated, as the circulation changes in the tropics accompanying global warming are unlikely to be the same as those during an El Niño (Mitchell, 1991; Boer, 1991).

Cloud feedback effects have been investigated in GCM studies in order to understand some of the consequences of the different ways in which clouds and cloud optical properties are represented in models. Clouds are represented either diagnostically as a function of relative humidity, or prognostically by carrying an equation for cloud water. Additionally, the optical properties of the clouds may be specified to remain fixed or they may be parametrized to change as the climate changes (see Table B1; also Table 3.2a, in IPCC 1990). As noted in IPCC 1990, Cess *et al.* (1990) included examples of each of these possibilities in a study of 19 GCMs, and found that the simulated cloud feedback varied from slightly negative to strongly positive.

Cloud feedback due to changes in cloud amount and/or distribution may be different, depending on whether clouds are specified in terms of relative humidity or derived from an explicit cloud water prognostic equation. Differences may also arise from the differing treatment of clouds in explicit cloud water schemes. For example, in the UKMO model (Senior and Mitchell, 1992a) increasing the lifetime of water cloud relative to ice cloud weakens the cloud feedback (tending to reduce climate sensitivity), while in the LMD model (Li and Le Treut, 1991) lowering the temperature at which ice cloud forms strengthens the cloud feedback. In the UKMO model the parametrization of cloud properties (as a function of cloud water content) apparently results in a modest positive contribution to the total feedback in the tropics due to increased infrared longwave emissivities. In the CCC model, on the other hand, the parametrization of cloud properties (as a function of temperature) gives a negative contribution to the feedback in tropical regions due to the increase in cloud albedo. The reduced tropical solar input at the surface in the perturbed climate thus moderates both the warming and the increase in evaporation, so that the hydrological

cycle does not strengthen as much as in other models (Boer, 1991).

B3.4 Surface Albedo Feedback

The conventional explanation of the amplification of global warming by snow feedback is that a warmer climate will have less snow cover, resulting in a darker surface which in turn absorbs more solar radiation. A recent analysis suggests that snow-albedo-temperature feedback processes in models are somewhat more complex than this view would indicate.

Using a methodology developed for an earlier study of GCM cloud feedbacks (Cess *et al.*, 1990), a perpetual April simulation was used with an imposed globally-uniform SST perturbation of 2°C relative to a prescribed climatological SST distribution. To isolate snow feedback, two such April climate change simulations were performed, one in which the snow cover was held fixed and one in which the snow line was allowed to retreat following the imposition of the SST anomaly. The results of these simulations are shown in Figure B18 in terms of the climate feedback or sensitivity parameter λ that was defined in Section 3.3.1 of the IPCC (1990) report. (An increase in λ represents an increased climate change caused by a given climate forcing.) There are clear differences among current GCMs in their response to changes in snow cover as measured by the ratio λ/λ_S (where λ_S is the sensitivity parameter for fixed snow). Here a value of $\lambda/\lambda_S > 1$ indicating a positive snow feedback is found in most models. Also shown are the corresponding feedbacks in the case of an equivalent cloudless atmosphere, found by separately averaging the models' clear-sky radiative fluxes at the top of the atmosphere (TOA) in order to isolate the effects of clouds on snow

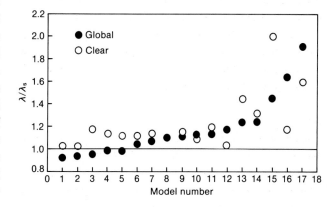

Figure B18: The snow feedback in terms of the ratio of the sensitivity parameter (λ) to that with fixed snow (λ_S), as simulated in 17 atmospheric GCMs. The full circles denote global values while the open circles are for clear-sky conditions. (From Cess *et al.*, 1991.)

feedback. The clear-sky values of λ/λ_s range from negligible snow feedback to as much as a two-fold feedback amplification.

In general, clouds are thought to reduce positive snow feedback by shielding the TOA albedo change. This mechanism, however, cannot account for the feedback sign reversals or the cloud-induced amplification of positive snow feedback exhibited by some models. As discussed by Cess *et al.* (1991), the sign reversal in some cases is caused by cloud redistribution, while in others it occurs as a consequence of cloud-induced longwave interactions; there is also a significant longwave feedback associated with snow retreat in some models. These results are complementary to those of Cohen and Rind (1991), who found that the suppression of surface air temperature changes by positive snow anomalies is considerably reduced by a negative feedback involving the non-radiative components of the surface energy budget.

Uncertainties in the size of the surface albedo feedback associated with sea-ice remain. As Ingram *et al.* (1989) noted, estimates of the strength of feedback depend not only on the model used, but also on the method used to provide the estimate. Covey *et al.* (1991) estimated an upper limit to sea-ice-albedo feedback by carrying out simulations in which the albedo of sea-ice was changed to that of the open ocean. They found an enhancement of globally and annually averaged absorption of radiation of 2 to $3Wm^{-2}$, comparable to the change on doubling CO_2. They concluded that for a warming of the magnitude expected on doubling CO_2, sea-ice-albedo feedback is likely to be smaller than the feedback from water vapour and potentially smaller than that from clouds. This is consistent with Ingram *et al.* (1989) who estimated a feedback of 0.2-$0.3Wm^{-2}K^{-1}$ from sea-ice-albedo changes, whereas estimates of the strength of water vapour feedback are typically about $1.5Wm^{-2}K^{-1}$ (see for example, Mitchell, 1989). Meehl and Washington (1990) found that changing their sea-ice albedo formulation to one which was liable to induce ice melt led to both a warmer control climate (about 1K) and a greater sensitivity (about 0.5K) to doubling CO_2. Further discussion of sea-ice modelling is given in Section B5.4.

B3.5 Climate Sensitivity
There is no compelling new evidence to warrant changing the equilibrium sensitivity to doubled CO_2 from the range of 1.5 to 4.5°C as given by IPCC 1990.

The climate sensitivity is defined as the equilibrium change in global average surface air temperature due to a doubling of CO_2 (Section 3.2 of IPCC, 1990), and is a measure of the response of a climate model to a change in radiative forcing. The climate sensitivity may be thought of as partly a direct radiative effect (estimated to be of the order of 1.2°C for a doubling of CO_2) and partly the effect

of feedbacks that act to enhance or suppress the radiative warming. The IPCC "best guess" for the climate sensitivity is 2.5°C, with a range of uncertainty from 1.5 to 4.5°C. Values of the climate sensitivity estimated from recent equilibrium GCM simulations for doubled CO_2 are summarized in Table B2, and generally fall within the range given in Table 3.2a of IPCC 1990. These recent simulations convey little new information as they all (with the exception of the LMD model) employ a relative humidity-based cloud scheme and fixed cloud radiative properties as in earlier models. The coupled model results (at the time of CO_2 doubling) give lower bounds, as discussed in Section B2.2.

Recent additional estimates of the climate sensitivity have been made by fitting the observed temperature record to the evolution of temperature produced by simple energy-balance climate/upwelling-diffusion ocean models, assuming that all the observed warming over the last century or so was due solely to increases in greenhouse gases (see IPCC (1990), Sections 6 and 8). Schlesinger *et al.* (1991) estimate a climate sensitivity of 2.2±0.8°C allowing for the effect of sulphate aerosols, while Raper *et al.* (1991) obtain a value of 2.3°C with a larger range of uncertainty due to their allowance for natural variability (see Wigley and Raper, 1990) and they estimate a value of 1.4°C with no aerosol effect. The effect of including sulphates (or other factors that could act to oppose the greenhouse warming) in such calculations is to increase the estimated climate sensitivity, since the observed climate warming is then ascribed to a reduced radiative forcing (see Sections A2.6 and C4.2.4).

In summary, there have been a number of further studies that have a bearing on estimates of climate sensitivity. New equilibrium GCM simulations have widened the range slightly to 1.7°C (Wang *et al.*, 1991a) and 5.4°C (Senior and Mitchell, 1992a), but no dramatically new sensitivity has been found. Energy-balance model considerations bring previous estimates of sensitivity (IPCC, 1990) more in line with the IPCC "best guess".

B4 Advances in Modelling Atmospheric Variability
1. Although studies are incomplete in many respects, the ability of models to replicate observed atmospheric behaviour on a wide range of space and time-scales is encouraging. As noted in IPCC 1990, this ability provides some evidence that models may be usefully applied to problems of climatic change associated with the greenhouse effect.

2. Model experiments with doubled CO_2 give no clear indication of a systematic change in the variability of temperature on daily to interannual time-scales, while the changes of variability for other climate features appear to be regionally (and possibly model) dependent.

B4.1 Introduction

Since the 1990 IPCC report there has been increased emphasis of the analysis of the ability of models to simulate atmospheric variability on time-scales ranging from diurnal to decadal in both control and perturbed climate simulations. Such studies are critical to the assessment of the effects of climate change, since it is increasingly recognized that changes in variability and the occurrence of extreme events may have a greater impact than changes in the mean climate itself (Katz and Brown, 1991). Notwithstanding these results, there is at present only limited confidence in the ability of climate models to infer changes in the occurrence of interannual and regional events.

The following discussion is arranged in order of increasing time-scales from diurnal to decadal. Particular attention is given to new studies that include simulated changes in variability due to doubling atmospheric CO_2, and to relating new results to the findings of IPCC 1990. Other studies are included to document the improvement in the capability of models to simulate selected aspects of atmospheric variability.

B4.2 Diurnal Cycle

Inclusion of the diurnal cycle (now a feature of many GCMs) permits the simulation of high-frequency temporal variations. For example, a recent simulation by Randall *et al.* (1991a) has demonstrated that the diurnal variation of the hydrologic cycle in the tropics can be realistically simulated, while Cao *et al.* (1992) have found that the simulated diurnal range of surface temperature is generally comparable to observations except for a slight underestimate in the mid-latitudes. On doubling CO_2, Cao *et al.* (1992) found a slight decrease in the globally-averaged diurnal range of surface temperatures, as noted in IPCC 1990, although locally they found that factors such as reduced soil moisture, receding snow lines or reduced cloud cover could lead to an increase in the diurnal temperature range with doubled CO_2. Historical observations over 25% of the global land area show a decreased diurnal temperature range, although the reasons for this change, which is largely a result of an increase in minimum temperatures, are not yet clear (see Section C3.1.5).

B4.3 Day-to-Day Variability

In several of the models used in CO_2 studies, the day-to-day variability of surface temperature is greater than that observed, particularly in higher northern latitudes (Meehl and Washington, 1990; Portman *et al.*, 1990). Cao *et al.* (1992) found that in a UKMO model the daily variability of surface air temperature is overestimated in high northern latitudes in winter, but is less than observed over mid-latitude continents in summer.

On doubling CO_2, Cao *et al.* (1992) found a general reduction in day-to-day temperature variability in winter in mid- to high latitudes over North America and over the ocean. The reduction was particularly pronounced in regions where winter sea-ice had melted. Mearns *et al.* (1990), however, found no clear pattern of change in daily temperature variability whereas P. Whetton (personal communication) found general increases in daily temperature variability throughout Australia in both winter and summer. Mearns *et al.* (1990) found a general increase in the day-to-day variability of precipitation due to increased CO_2, whereas Boer *et al.* (1991b) found an overall decline in the daily variance of sea-level pressure due to doubling CO_2, with the largest decrease occurring in the North Atlantic in winter.

B4.4 Extreme Events

The simulation of extreme events is an important aspect of a model's performance, and is closely connected with the question of natural variability. An important extreme event is the occurrence of tropical cyclones (or hurricanes) which GCMs cannot simulate in detail, though they do simulate tropical disturbances (see IPCC 1990, Section 5.3.3). Broccoli and Manabe (1990) and Haarsma *et al.* (1992) found that the spatial and temporal distributions of modelled tropical disturbances are similar to those observed. On doubling CO_2 Haarsma *et al.* (1992) found that the number of simulated tropical disturbances increased, with little change in their average structure and intensity. However, as noted in IPCC 1990, Broccoli and Manabe (1990) found an increase in the number of tropical storms if cloud cover was prescribed, but a decrease if cloud was generated within the model.

In the 1990 IPCC report a consistent increase in the frequency of convective precipitation at the expense of large-scale precipitation was noted, with the implication of more intense local rain at the expense of gentler but more persistent rainfall events. This tendency has been found in recent simulations using the CSIRO model (Gordon *et al.*, 1991; Pittock *et al.*, 1991) and a high-resolution UKMO model (J. Gregory, personal communication). Pittock *et al.* (1991) find a systematic increase in the frequency of heavy rain events with doubled CO_2 (Figure B19) and a consequent decrease in the return period of heavy rainfall (Figure B20). These changes are related to a systematic increase in the frequency of penetrating convection in the tropics and mid-latitudes on doubling CO_2 (see IPCC, 1990).

B4.5 Blocking and Storm Activity

Studies of the simulated variation of storm activity with enhanced CO_2 are difficult to generalize as they use different measures for storminess and address different regions. R. Lambert (personal communication) found a

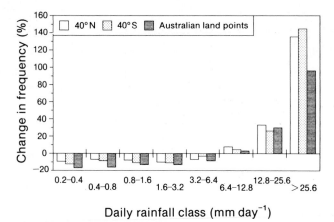

Figure B19: Changes in the frequency of occurrence of daily rainfall classes with doubled CO_2 in the CSIRO model. (From Pittock *et al.*, 1991.)

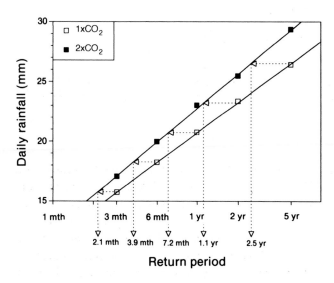

Figure B20: Daily rainfall amount vs. return period in Australia as simulated by the CSIRO model for both control ($1\times CO_2$) and doubled CO_2. (From Pittock *et al.*, 1991.)

significant reduction in the number of cyclonic events in the CCC model between 30°N and the North Pole during winter, but a significant increase in the number of strong cyclones. (Boer *et al.* (1991b) noted a decrease in the variance of 1000 hPa height in the same experiment; see Section B4.3.) Senior and Mitchell (1992b) found a northwest shift in the filtered variance of geopotential height (a surrogate for storm tracks) in the North Atlantic in winter in the high resolution UKMO model. B. Hoskins (personal communication), on the basis of experiments using the time-mean data from Senior and Mitchell's simulations, attributed the changes in storm tracks to changes in the horizontal temperature gradient and increases in atmospheric water vapour. Mullan and

Renwick (1990) conducted a detailed analysis of the results of the CSIRO model in the Australian-New Zealand region, and found a slight increase in the number of storms in this region for all seasons.

The simulation of persistent large-scale anomalies such as blocking has proved difficult to forecast with extended-range NWP models (Tibaldi and Molteni, 1990; Miyakoda and Sirutis, 1990), although GCMs can simulate some of the statistical properties of blocking in the Northern Hemisphere as recently shown by Hansen and Sutera (1990) for the NCAR CCM and by Kitoh (1989) for the MRI GCM. Tibaldi (1992) has examined the space-time variability of blocking in the MPI (ECHAM) model, using the blocking index of Tibaldi and Molteni (1990) applied to 5-day mean December to February 500 hPa fields in the Northern Hemisphere. While the variance of both the low- and high-frequency components is reasonably well simulated, the magnitudes are systematically underestimated.

B4.6 Intra-Seasonal Variability

An important intra-seasonal variability is that associated with the 30-60 day oscillation. This oscillation appears to influence onset and break phases of the monsoons, and thus has a considerable impact on the prediction of tropical rainfall (Lau and Peng, 1990). These oscillations may also play a role in the onset of El Niño events. Park *et al.* (1990) and Zeng *et al.* (1990) have shown that some GCMs simulate intra-seasonal variations that are similar in structure to those observed, but they are generally of smaller amplitude.

B4.7 Interannual Variability

As in IPCC 1990, no meaningful change in the interannual variability of temperature has been found with increased CO_2, apart from a general reduction in the vicinity of the winter sea-ice margins.

Some of the models used in climate studies exaggerate the interannual variability of surface temperature in high latitudes in winter (Mearns, 1991, reporting on simulations by Oglesby and Saltzman, 1990; Cao *et al.*, 1992). The simulated changes in the variability of temperature with doubled CO_2 vary from model to model; Rind (1991) found decreases over much of the northern continents in winter in the GISS model, as did Georgi *et al.* (1991) using a regional model driven by output from the NCAR GCM simulations of Washington and Meehl (1991). On the other hand, Mearns (1991) found both increases and decreases in temperature variability, while Cao *et al.* (1992) found that increases were widespread in the tropics and subtropics, and in summer over North America and western Europe, with significant reductions confined to regions where sea-ice receded in winter.

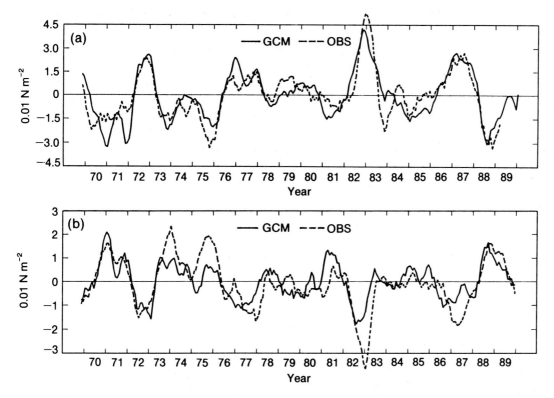

Figure B21: The 1970-1989 time-series of (a) the zonal surface wind stress anomalies averaged between 180°-140°W and 4°N-4°S, and (b) the Southern Oscillation Index. The observed variation is given by the dashed line, and that simulated by the MRI GCM is given by the solid line. (From Kitoh, 1991.)

B4.8 ENSO and Monsoons

A dominant mode of atmospheric variability in the tropics is that associated with tropical sea surface temperature anomalies in the Pacific. These tropical ocean-global atmosphere oscillations are called El Niño-Southern Oscillation, or ENSO. As noted in IPCC (1990), atmospheric models are capable of giving a realistic simulation of the seasonal tropical atmospheric anomalies at least for intense El Niño periods if they are given a satisfactory estimate of the anomalous sea surface temperature (SST) in the tropical Pacific. This conclusion is supported by more recent results (König *et al.*, 1990; Kitoh, 1991; B. Hunt, personal communication). For example, as illustrated in Figure B21, a rather good simulation has been made of the observed interannual variations of zonal surface wind stress over the central equatorial Pacific and of the observed large-scale interannual variations of sea-level pressure as represented by the Southern Oscillation Index. Using prescribed sea surface temperatures from 1987 and 1988, the broad characteristics of monsoon interannual variability have been reproduced (Palmer *et al.*, 1991; Kitoh, 1992). Using the coupled IAP model with observed initial conditions, Zeng *et al.* (1990) and Li *et al.* (1991) report a successful simulation of summer precipitation anomalies in the monsoon region of southeast Asia.

Using tropical ocean-atmosphere models, Lau *et al.* (1991), Nagai *et al.* (1991), Philander *et al.* (1991) and Latif *et al.* (1991) have successfully simulated interannual variations that resemble some aspects of ENSO phenomena, although the amplitude of the simulated SST anomalies is generally less than that observed (as noted in IPCC, 1990) and model resolution appears to have an important influence on the simulated behaviour. Meehl (1990) claims some success in simulating ENSO-like disturbances in a global low-resolution ocean-atmosphere GCM, though it should be recalled that not all tropical SST variations are associated with ENSO. As shown in Figure B22, the patterns of tropical sea-level pressure changes that are characteristic of ENSO in the present climate are also present with doubled CO_2 (Meehl *et al.*, 1991b). The patterns of the tropical precipitation and soil-moisture anomalies are similar in the control and doubled CO_2 cases, with the dry areas becoming generally drier and the wet areas becoming generally wetter with increased CO_2. These results, however, are yet to be confirmed with other models. On increasing the concentration of greenhouse gases in the MPI global coupled model, Lal *et al.* (1992) found no evidence for a significant change in the mean onset date of the Indian Monsoon or for changes in the precipitation in the monsoon region.

(a)

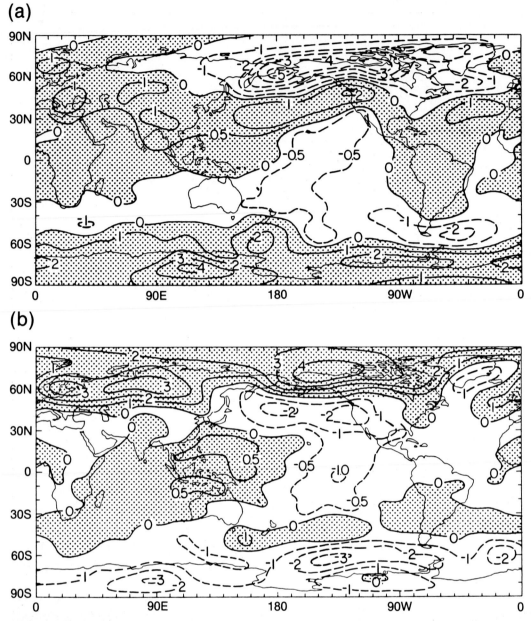

(b)

Figure B22: The change in DJF sea-level pressure (hPa) between composite warm ENSO events and the 15-year mean simulated by a coupled ocean-atmosphere model for (a) normal CO_2, and (b) doubled CO_2. (From Meehl *et al.*, 1991b.)

B4.9 Decadal Variability

There are now a sufficient number of model integrations over 50-100 year (and longer) periods to provide preliminary information on the simulation of atmospheric decadal variability. (See Section B4.3 for a discussion of simulated decadal variability in the ocean.) Whether simple models, low-resolution GCMs coupled to an oceanic mixed layer (Houghton *et al.*, 1991; B. Hunt, personal communication) or full global ocean-atmosphere GCMs (Section B2.2) are used, the presence of considerable natural variability (i.e., in the absence of changes in external forcing) on decadal time-scales is characteristically found in model control runs. Typical magnitudes of such decadal variations are several tenths °C in surface air temperature, as illustrated in Figure B2 for

the control run of the GFDL coupled model that was used in the transient CO_2 simulation discussed in Section B2.2. In this integration the surface flux corrections applied at the ocean surface were evidently successful in preventing a multi-decadal drift of the mean temperature. Such internal variability may mask at least part of the changes due to increased greenhouse gases.

B5 Advances in Modelling the Oceans and Sea-Ice

B5.1 Introduction

Although there is generally less experience in ocean modelling than there is in atmospheric modelling, since the preparation of the IPCC (1990) report there have been a number of studies with high-resolution and other ocean

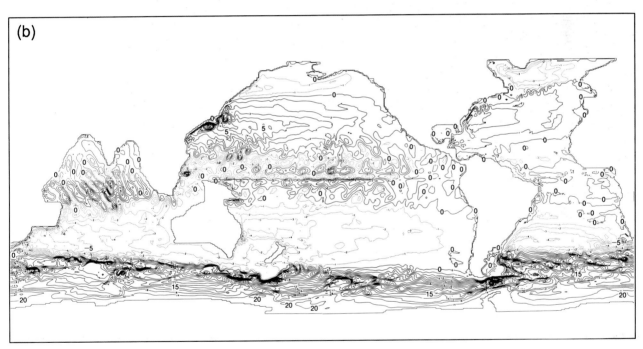

Figure B23: (a) The instantaneous 160m temperature at 1°C intervals, and (b) the volume transport streamlines at 10 Sv intervals, as simulated in an eddy-resolving global ocean GCM with 1/2° resolution. (From Semtner and Chervin, 1988.)

models that have potentially important consequences for the simulation (and identification) of climate change with coupled ocean-atmosphere GCMs (Anderson and Willebrand, 1991).

B5.2 Eddy-Resolving Models

Recent simulations with the WOCE community ocean model (Bryan and Holland, 1989) for the North Atlantic with a 1/3° resolution have yielded a distribution of eddy energy that agrees reasonably well with the GEOSAT altimetric observations, although the simulated amplitudes are generally too small. A qualitatively similar agreement of simulated and observed variability was found by Semtner and Chervin (1988) with a 1/2° model of the global ocean, examples of which are shown in Figure B23. First results from the FRAM 1/4° model for the Antarctic Circumpolar Current (Webb *et al.*, 1991) indicate that the distribution of maximum variability associated with mesoscale eddies is reasonably well simulated, although the eddy minima appear to be too low. The FRAM model also shows a strong meridional (Deacon) cell in the Antarctic Ocean, with a transport of over 20Sv. The upwelling branch of this cell, which is seen in both high- and low-resolution models, occurs at those latitudes where the coupled ocean-atmosphere models show delayed greenhouse gas warming (see Section B2.2). A recent experiment by Böning and Budich (1991) has shown that the simulated eddy energy increases substantially (along with significant variability on time-scales of several years) when the horizontal resolution is increased to 1/6°. This raises the question (of practical importance for climate modelling) whether or not mesoscale ocean eddies must be resolved in climate calculations. Mesoscale effects are so inhomogeneous and so tenuously connected to the larger-scale currents that a simple parametrization of their effects seems unlikely. However, studies using models that omit salinity variations suggest that the total meridional heat transport is only slightly influenced by eddies, since the eddy flux tends to be compensated by a modification of the mean flow induced by the eddies themselves (Bryan, 1991). Further research on the climatic (as opposed to the synoptic) role of ocean eddies is clearly required.

B5.3 Thermohaline Circulation

A great deal of interest has recently focussed on the behaviour of the thermohaline circulation of the North Atlantic, whereby North Atlantic deep water is formed. This sinking is part of a global system of ocean transports known as the Conveyor Belt Circulation (Gordon, 1986). Many aspects of the thermohaline circulation have been simulated in a 1/2° global ocean model (Semtner and Chervin, 1991), and its basic dynamical structure and sensitivity are being illuminated by both theoretical studies and simulations with idealized models. In the present climate, high-latitude winter cooling of relatively saline surface water in the North Atlantic causes it to sink and subsequently to move out of the basin by deep southward currents; this flow is compensated by the northward flow of warm surface water from the tropics, whose salinity may be increased relative to the other high-latitude oceans by enhanced evaporation (Stocker and Wright, 1991).

Coupled ocean-atmosphere models produce sinking in high latitudes in the North Atlantic, though it is not clear how realistically they do so. It may also be noted that Dixon et al. (1991) have recently succeeded in simulating the observed spreading of CFCs in the Southern Ocean with the same GFDL coupled model that was used in the transient CO_2 experiments; this increases confidence that the vertical mixing parametrization in the oceanic part of the model is a reasonable approximation. However, it should be noted that heat may affect the vertical stability in the ocean and hence may not behave in the same way as

CFCs. Using a model with flux adjustments, Manabe et al. (1991) find that the effect of enhanced greenhouse warming is to reduce the rate of deep water formation. (Washington and Meehl (1989) also note a reduction in deep water formation in a model without flux adjustments.) This in turn decreases the flow of warm surface water from the south and probably contributes to a delay in the greenhouse response in the northern North Atlantic noted earlier in Figure B4.

As shown with ocean-only models, variability on decadal and longer time-scales is associated with the advection of salinity anomalies through the deep convection regions (Marotzke and Willebrand, 1991; Weaver and Sarachik, 1991; Weaver et al., 1991; Mikolajewicz and Maier-Reimer, 1990). In the GFDL coupled model (R. Stouffer, personal communication) similar variability on time-scales of order 50 years is found. In an integration started from different initial conditions, a GFDL model (Manabe and Stouffer, 1988) in fact converged to a second equilibrium solution in which the North Atlantic was both fresher and colder (as shown in Figure B24) with a thermohaline circulation that was weakly reversed. The existence of multiple equilibrium states for the global ocean circulation raises the question of their stability, i.e., how easily the system may change from one state to another. Maier-Reimer and Mikolajewicz (1989) used the uncoupled MPI ocean GCM to investigate the Younger-Dryas event and found that the North Atlantic thermohaline circulation was sensitive to relatively small variations in the strength and location of the surface fresh water input, such that a breakdown of the circulation could occur within a few decades. Other recent experiments with zonally averaged ocean models have demonstrated that transitions between multiple equilibrium states can be triggered by relatively modest changes in the large-scale precipitation (Marotzke and Willebrand, 1991; Stocker and Wright, 1991).

The possibility of the natural collapse of the North Atlantic thermohaline circulation has obvious implications for our ability to predict future variations of the climate system on decadal and longer time-scales. The North Atlantic circulation, however, appears to be more robust to change than the ocean-only model results suggest. Atlantic and Pacific deep sea sediment cores show that, with the exception of a possible short interruption during the Younger-Dryas event, no major break in the thermohaline circulation has occurred since the end of the last ice age in spite of the variability that has occurred in both the atmosphere and ocean (see, for example, Keigwin et al., 1991). Preliminary results from a long integration of a coupled ocean-atmosphere model (R. Stouffer, personal communication) suggest that the thermohaline circulation is more stable in a coupled model than in an ocean-only model with restored SSTs.

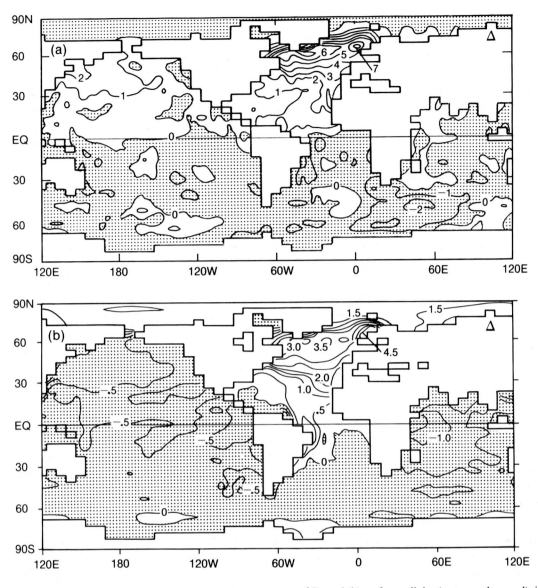

Figure B24: The difference in (a) mean surface temperature (°C), and (b) surface salinity (parts per thousand), in the North Atlantic in an equilibrium solution of the GFDL coupled ocean-atmosphere model with an active thermohaline circulation, relative to an alternative solution without thermohaline circulation. (From Manabe and Stouffer, 1988.)

B5.4 Sea-Ice Models

Ocean-only and coupled ocean-atmosphere models generally use thermodynamic sea-ice models with (at most) only simple advection schemes for the sea-ice. Experiments with sea-ice models coupled to mixed-layer pycnocline models (Owens and Lemke, 1990) have shown that the net freezing rate may depend on the continuum-mechanical properties used for the sea-ice. Inclusion of sea-ice dynamics may therefore be required to properly model the surface boundary conditions of the ocean. A model that contains most of the characteristics of sea-ice believed to be relevant for climate investigations was proposed by Hibler (1979) and has recently been reformulated in a simplified version by Flato and Hibler (1990).

Experiments with dynamic-thermodynamic sea-ice models have shown that such models are generally less sensitive to changes in the thermal forcing than are the simplified sea-ice models used so far in climate studies (Hibler, 1984; Lemke *et al.*, 1990). This reduced sensitivity is a result of a negative feedback between the sea-ice dynamics and thermodynamics. It is therefore possible that the pronounced response of the polar regions found in experiments with increased CO_2 (see Section B2.2 and IPCC (1990), Section 5) will be modified by the inclusion of ice dynamics. On the other hand, dynamic sea-ice models are more sensitive than thermodynamic models to changes in the wind forcing, and most atmospheric circulation models do not reproduce the observed wind field in polar regions very well. Further improvement of sea-ice models therefore depends upon advances in the parametrization of polar processes, including the effect of

fog and low stratus clouds on the polar radiation balance. It
may also be noted that models typically filter their
solutions in polar latitudes for purely numerical reasons,
and this may tend to degrade the simulation of sea-ice.

B6 Advances in Model Validation

B6.1 Introduction

Climate models show an increasing ability to simulate the
current climate and its variations. This improvement is due
to a combination of increased model resolution and
improved physical parametrizations. The correctness of
simulated results for the present climate at both global and
regional scales is a desirable but not a sufficient condition
for increased confidence in their use for simulations of
future climate change.

B6.2 Systematic Errors and Model Intercomparison

Every climate modelling group strives to identify their
models' systematic errors as a natural part of the
continuing process of model development. Ideally, each
physical parametrization in a climate model should be
individually calibrated against appropriate observational
data, although in many (if not most) cases this is not
possible. Instead, with a particular set of parametrizations
and a particular resolution, a model's performance is
evaluated against the available seasonal climatological
distributions of the large-scale variables such as
temperature and circulation. While atmospheric and
oceanic models continue to improve, the overall
characterization of their performance given in Section 4 of
IPCC (1990) remains valid (Gates *et al.*, 1990). The
situation is somewhat different, however, for coupled
atmosphere-ocean GCMs, for which only a provisional
error assessment is possible due to the limited number of
integrations that have been performed (see below and
Section B2.2) and because of the limited ocean data
currently available.

Progress in atmospheric model validation since the
IPCC (1990) report has occurred both in the form of
documented improvements in specific GCMs and in the
form of intercomparisons of large numbers of independent
GCMs. An example of an improvement in a particular
model is the change in the parametrization of tropical
evaporation and convection that has removed a major
systematic error in the ECMWF model (see Figure B25).

Recent intercomparisons of atmospheric models have
either focussed on the simulation of the present climate (as
in Boer *et al.*, 1991a) or have considered the radiative
forcing induced by a prescribed climate change (as in Cess
et al., 1991) (see Section B3.3.3). On behalf of the
Working Group on Numerical Experimentation (WGNE),
Boer *et al.* (1991a) have collected the mean DJF and JJA
distributions of selected atmospheric variables as

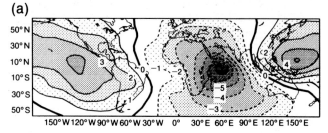

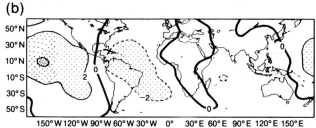

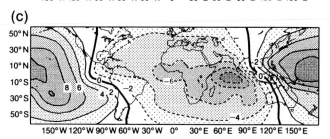

Figure B25: The 200 hPa velocity potential anomalies ($10^6 m^2 s^{-1}$)
for JJA 1988: (a) as given by ECMWF analyses referenced
against the mean during 1986-1990; (b) as simulated by the
ECMWF model over the same interval with observed SST and
standard surface flux parametrizations for heat and moisture, and
(c) as simulated with revised parametrizations accounting for
fluxes at low windspeeds. (From Miller *et al.*, 1991.)

simulated in the control runs of 14 atmospheric GCMs.
Despite the fact that the models differed greatly in
resolution and in the nature and sophistication of their
parametrizations of physical processes, a number of
common systematic errors where found. The simulated
climate was colder than that observed on average, and all
models showed a cold bias in the middle- and high-latitude
upper troposphere and in the tropical lower stratosphere
(see Figure B26). These temperature errors induce
corresponding errors in the zonal wind distribution in
accordance with the thermal wind relationship, with the
result that most GCMs display too westerly a flow in the
upper troposphere and in high latitudes. Taken as a whole,
models appear to be more sensitive to changes in physical
parametrizations than to modest changes in resolution.

The principal features of the Hadley, Ferrel and polar
cells that characterize the atmospheric circulation are
simulated by all models, along with appropriate seasonal
shifts. There is, however, considerable variation in the
strengths of the models' simulated meridional circulations.

(a)

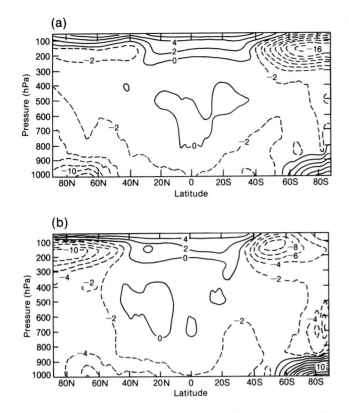

(b)

Figure B26: The systematic error (simulated minus observed) in mean temperature (˚C) for (a) DJF, and (b) JJA for the UKMO atmospheric GCM. (From Boer *et al.*, 1991a.)

At the same time there has been improvement in the simulation of mean sea-level pressure in a number of models, due partly to increased resolution and partly to the introduction of the parametrization of gravity wave drag. This improvement is illustrated in Figure B27a. The largest differences among models are found near Antarctica, where the extrapolation of pressure to sea level may introduce artificial variations.

There has not been as much improvement in the simulation of precipitation by current GCMs. As shown in Figure B27b, there is a systematic over-estimation of the precipitation over much of the Northern Hemisphere (as well as an apparent under-estimation in the mid-latitudes of the Southern Hemisphere), and the models' simulated precipitation in the tropics varies considerably around that observed. There are, however, uncertainties in the observed data themselves, especially over the tropical oceans and in the Southern Hemisphere. Relatively large uncertainties in the simulation of regional precipitation are shown in the intercomparison of atmospheric GCMs being undertaken by the Monsoon Numerical Experimentation Group (MONEG) of the WCRP (1990b).

In a companion study to the model intercomparisons of Cess *et al.* (1990, 1991) (see Section B3.3), an intercomparison of the surface energy fluxes in the GCMs has recently been made (Randall *et al.*, 1991b). In this study

(a) JJA

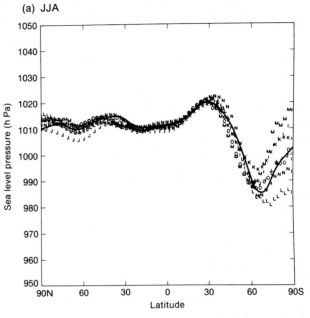

(b) DJF

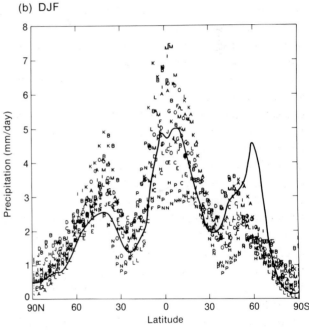

Figure B27: (a) The zonally-averaged JJA sea level pressure (mb) simulated by current high resolution GCMs; (b) the zonally-averaged DJF precipitation (mm/day) simulated by current GCMs. Different symbols indicate different models and the solid line is the observed climatological average according to Jaeger (1976). (From Boer *et al.*, 1991a.)

considerable variation is found in the models' simulated increases in precipitation, evaporation and total atmospheric water vapour in response to a uniform 4˚C SST increase. As shown in Figure B28, the models also simulate an overall increase in the net infrared clear-sky flux at the surface, reflecting the positive temperature/water vapour feedback (Section B3.2) in the models.

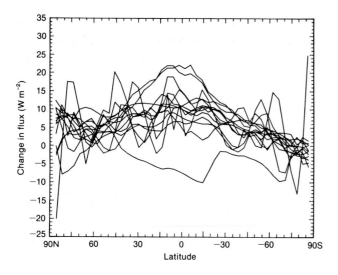

Figure B28: The change in the zonally-averaged longwave clear-sky flux at the surface as simulated by 14 atmospheric GCMs in response to a uniform 4°C increase of SST relative to July climatology. (From Randall *et al.*, 1991b.)

In the Boer *et al.* (1991a) intercomparison, as in most previous intercomparisons of climate models, results have been collected as available from the modelling community and no attempt has been made to run the models under common conditions and for common periods of time. In an effort to achieve a more systematic intercomparison of models, these attributes have recently been incorporated into the WGNE/PCMDI Atmospheric Model Intercomparison Project (AMIP), in which virtually all of the world's atmospheric GCMs will simulate the decade 1979-1988 with common values of the CO_2 concentration and solar constant and with observationally-prescribed but monthly-varying SST and sea-ice distributions (Gates, 1991). The AMIP involves many of the same modelling groups that participated in the earlier WCRP Intercomparison of Radiation Codes in Climate Models (ICRCCM), the results of which have recently been published (Ellingson *et al.*, 1991; Fouquart *et al.*, 1991).

The identification of systematic errors and the intercomparison of ocean models are at a less advanced stage than for atmospheric models. When forced with "observed" surface temperatures, salinities and wind stresses, ocean models have been moderately successful in reproducing the observed large-scale circulation and water mass distribution (see Section B5). The principal systematic errors common to most ocean models are an underestimate of the meridional heat transport as inferred from observations, and the simulation of the main thermocline to be too deep and too diffuse. No common systematic errors have as yet been identified in the deeper ocean in view of the paucity of observational data, although there are considerable differences in the

simulated rate and location of deep water formation and in the strength of deep ocean circulation.

In order to examine ocean model performance in more detail, an intercomparison of tropical Pacific ocean models has recently been undertaken by TOGA-NEG (WCRP, 1990a; D. Anderson, personal communication), in which several models have been integrated with the same representation of surface forcing. In general, the SST error in all models shows a similar pattern, with the central equatorial Pacific being too cold and the equatorial western Pacific too warm throughout the year. Water along the western coast of South America is also generally too warm, and the equatorial undercurrent tends to be too weak. These discrepancies may be at least partly caused by inadequate surface forcing.

An intercomparison of the behaviour of the tropical Pacific SST in coupled ocean-atmosphere models (in which the ocean and/or the atmosphere is a GCM) has recently been made by Neelin *et al.* (1992). These models show a wide range of behaviour in their simulation of tropical interannual variability, and some are more successful in simulating ENSO-like events than others. The runs were not made under the same conditions, however, and there is a wide range of model sophistication represented.

The comparison of simulated past climates with palaeo data provides a further test of models, though the usefulness of such tests is limited by the quality of data and by our knowledge of the changes of past climates. A Palaeoclimate Model Intercomparison Project (PMIP) has been proposed (NATO Advanced Research Workshop on Palaeoclimatic Modelling, 27-31 May 1991, Saclay, France) to promote the understanding of the response of climate models to past changes in forcing.

Although model intercomparison of the sort described here does not address many important questions in climate modelling (such as how to determine which parametrizations are "best" or how best to couple the atmosphere and ocean), intercomparison has proven useful as a collective exploration of model parameter space and as a benchmark for the documentation of model improvement. It is characteristic of model intercomparisons that no single model is found to be superior to all other models in all respects. It should also be recognized that the various intercomparisons are designed for different purposes: those of Cess *et al.* (1990, 1991) are concerned with the models' response to changes in forcing (and are therefore relevant to the response to possible future changes of greenhouse gases), while those of Boer *et al.* (1991a) and those being undertaken in AMIP (Gates, 1991) are concerned with the models' ability to simulate the present climate. As model development continues (and as increased computer resources become available), there will be both an opportunity and a need to conduct further climate model intercomparisons, especially of coupled ocean-atmosphere models in both their control configuration and in experiments with increasing greenhouse gases.

B6.3 Data for Model Validation

The availability of appropriate observational data is a critical factor in the validation (and improvement) of climate models, and some progress has been made in the assembly of global data sets for selected climate variables. The recent global compilations of average monthly surface air temperature and precipitation by Legates and Willmott (1990a, b) are believed to be improvements over earlier atlases of these variables, and new assemblies of land-based precipitation (Hulme, 1992) and soil moisture (Vinnikov and Yeserkepova, 1991) are available. It should also be noted that the diagnostics made from the operational analyses of global numerical weather prediction models (Hoskins *et al.*, 1989; Trenberth and Olson, 1988) are important sources of data for the validation of selected aspects of atmospheric models. The climatological ocean atlas of Levitus (1982) has been supplemented by the compilation of surface variables from the COADS data set (Wright, 1988; Oberhuber, 1988; Michaud and Lin, 1991). These data are proving useful in the validation of both ocean and coupled ocean-atmosphere models, as are the observations of transient tracers in the ocean (Toggweiler *et al.*, 1989; Dixon *et al.*, 1991).

For other variables such as cloudiness, precipitation, evaporation, run-off, surface heat flux, surface stress, ocean currents and sea-ice, however, the observational data base remains inadequate for the purposes of model validation. These variables are relatively difficult to observe on a global basis, and are not easily inferred from compilations of conventional circulation statistics such as those of Oort (1983). The best prospect for their systematic global estimation is probably the procedure known as re-analysis, whereby modern data assimilation techniques are used to retrospectively initialize a comprehensive atmospheric GCM on a daily basis for a number of past years. Such projects are being planned by both ECMWF and NMC, and the possibility of applying the technique to ocean models is under active consideration.

Finally, it should be noted that the development of a more comprehensive global climate data base is a key element of the World Climate Research Programme (WCRP). The observational plans that have been developed in recent years for the Global Precipitation Climatology Project (WCRP, 1988a), for the World Ocean Circulation Experiment WOCE (WCRP, 1988b), for the Tropical Ocean-Global Atmosphere project TOGA (WCRP, 1990a), and for the Global Energy and Water Cycle Experiment GEWEX (WCRP, 1990c) are focussed on the acquisition of data that are necessary for the further development and validation of global atmospheric and oceanic models, and have culminated in the proposal for a Global Climate Observing System (GCOS) (WCRP, 1991).

References

Anderson, D.L.T. and J. Willebrand, 1991: Recent advances in modelling the ocean circulation and its effects on climate. *Rev. Progress Phys.* (In press).

Arking, A., 1991: Temperature variability and water vapor feedback. (Unpublished manuscript).

Betts, A.K. and Harshvardhan, 1987: Thermodynamic constraint on the cloud liquid water feedback in climate models. *J. Geophys. Res.*, **92**, 8483-8485.

Betts, A.K., 1990: Greenhouse warming and the tropical water budget. *Bull. Amer. Met. Soc.*, **71**, 1464-1465.

Boer, G., 1991: Climate change and the regulation of the surface moisture and energy budgets. (Unpublished manuscript).

Boer, G.J., K. Arpe, M. Blackburn, M. Déqué, W.L. Gates, T.L. Hart, H. Le Treut, E. Roeckner, D.A. Sheinin, I. Simmonds, R.N.B. Smith, T. Tokioka, R.T. Wetherald and D. Williamson, 1991a: *An intercomparison of the climates simulated by 14 atmospheric general circulation models.* CAS/JSC Working Group on Numerical Experimentation Report No.15, WMO/TD - No. 425, World Met. Organiz., Geneva, 37 pp.

Boer, G.J., N.A. McFarlane and M. Lazare, 1991b: Greenhouse gas induced climate change simulated with the Canadian Climate Centre second generation general circulation model. *J. Climate* (In press).

Böning, C.W. and R. Budich, 1991: Eddy dynamics in primitive equation models: Sensitivity to horizontal resolution and friction. (Unpublished manuscript).

Broccoli, A.J. and S. Manabe, 1990: Can existing climate models be used to study anthropogenic changes in tropical cyclone climate? *Geophys. Res. Letters*, **17**, 1917-1920.

Bryan, F. and W.R. Holland, 1989: A high-resolution simulation of the wind- and thermohaline-driven circulation in the North Atlantic Ocean: In *Parametrization of Small-scale Processes*, Proceedings of 'Aha Huliko'a Hawaiian Winter Workshop, Univ. Hawaii, Manoa, pp. 17-20, 99-115.

Bryan, K., 1991: Poleward heat transport in the ocean: A review of a hierarchy of models of increasing resolution. *Tellus*, **43** A-B, 104-115.

Cao, H-X., J.F.B. Mitchell and J.R. Lavery, 1992: Simulated diurnal range and variability of surface temperature in a global climate model for present and doubled CO_2. *J. Climate*. (In press).

Cess, R.D., G.L. Potter, J.P. Blanchet, G.J. Boer, A.D. Del Genio, M. Déqué, V. Dymnikov, V. Galin, W.L. Gates, S.J. Ghan, J.T. Kiehl, A.A. Lacis, H. Le Treut, Z.-X. Li, X.-Z. Liang, B.J. McAvaney, V.P. Meleshko, J.F.B. Mitchell, J.-J. Morcrette, D.A. Randall, L. Rikus, E. Roeckner, J.F. Royer, U. Schlese, D.A. Sheinin, A. Slingo, A.P. Sokolov, K.E. Taylor, W.M. Washington, R.T. Wetherald, I. Yagai and M.-H. Zhang,

1990: Intercomparison and interpretation of climate feedback processes in nineteen atmospheric general circulation models. *J. Geophys. Res.*, **95**, 16,601-16,615.

Cess, R.D., G.L. Potter, M.-H. Zhang, J.-P. Blanchet, S. Chalita, R. Colman, D.A. Dazlich, A.D. Del Genio, V. Dymnikov, V. Galin, D. Jerrett, E. Keup, A.A. Lacis, H. Le Treut, X.-Z. Liang, J.-F. Mahfouf, B.J. McAvaney, V.P. Meleshko, J.F.B. Mitchell, J.-J. Morcrette, P.M. Norris, D.A. Randall, L. Rikus, E. Roeckner, J.-F. Royer, U. Schlese, D.A. Sheinin, J.M. Slingo, A.P. Sokolov, K.E. Taylor, W.M. Washington, R.T. Wetherald and I. Yagai, 1991: Interpretation of snow-climate feedback as produced by 17 general circulation models. *Science*, **253**, 888-892.

Charlson, R.J., J. Langer, H. Rodhe, C.B. Leovy and S.G. Warren, 1991: Perturbation of the Northern Hemisphere radiative balance by backscattering from anthropogenic sulfate aerosols. *Tellus*, **43AB**, 152-163.

Cohen, J. and D. Rind, 1991: The effect of snow cover on the climate. *J. Climate*, **4**, 689-706.

Covey, C., K.E. Taylor and R.E. Dickinson, 1991: Upper limit for sea-ice albedo feedback contribution to global warming. *J. Geophys. Res.*, **96**, 9169-9174.

Cubasch, U., K. Hasselmann, H. Höck, E. Maier-Reimer, U. Mikolajewicz, B.D. Santer and R. Sausen, 1991: *Time-dependent greenhouse warming computations with a coupled ocean-atmosphere model.* Max Planck Inst. Meteor. Report 67, Hamburg.

Del Genio, A.D., A.A. Lacis and R.A. Ruedy, 1991: Simulations of the effect of a warmer climate on atmospheric humidity. *Nature*, **251**, 382-385.

Dix, M., B.J. McAvaney, I.G. Watterson and H.B. Gordon, 1991: Personnal communication. Full text in preparation.

Dixon, K.W., J.L. Bullister, R.H. Gammon, R.J. Stouffer and G.P.J. Theile, 1991: Climate model simulation studies using chlorofluorocarbons as transient oceanic tracers. (Unpublished manuscript).

Ellingson, R.G., J. Ellis and S. Fels, 1991: The intercomparison of radiation codes used in climate models: Longwave results. *J. Geophys. Res.*, **96**, 8929-8954.

Feigelson, E.M., 1978: Preliminary radiation model of a cloudy atmosphere. Structure of clouds and solar radiation. *Contrib. Atmos. Phys.*, **51**, 203-229.

Flato, G.M. and W.D. Hibler, 1990: On a simple sea-ice dynamics model for climate studies. *Ann. Glaciol.*, **14**, 72-77.

Fouquart, Y., B. Bonnel and V. Ramaswamy, 1991: Inter-comparing shortwave radiation codes for climate studies. *J. Geophys. Res.*, **96**, 8955-8969.

Gates, W.L., 1991: *The WGNE atmospheric model inter-comparison project.* AMIP Newsletter **No. 1**, PCMDI, Lawrence Livermore Nat'l. Lab., Livermore, USA. 8pp.

Gates, W.L., P.R. Rowntree and Q.-C. Zeng, 1990: Validation of climate models. In: *Climate Change: The IPCC Scientific Assessment.* J.T. Houghton, G.J. Jenkins and J.J. Ephraums (Eds.). Cambridge University Press, Cambridge. pp93-130.

Giorgi, F. and L.O. Mearns, 1991: Approaches to the simulation of regional climate change: A review. *J. Geophys. Res.*, **29**, 191-216.

Giorgi, F., M.R. Marinucci and G. Visconti, 1990: Use of a limited-area model nested in a general circulation model for regional climate simulation over Europe. *J. Geophys. Res.*, **95**, 18,413-18,431.

Giorgi, F., M.R. Marinucci and G. Visconti, 1991: A $2\times CO_2$ climate change scenario over Europe generated using a limited-area model nested in a general circulation model. Part II. Climate change scenario. (Unpublished manuscript).

Gordon, A.L., 1986: Inter-ocean exchange of thermocline water. *J. Geophys. Res.*, **91**, 5037-5046.

Gordon, H.B., P. Whetton, A.B. Pittock, A. Fowler, M. Haylock and K. Hennessy, 1991: Simulated changes in daily rainfall intensity due to the enhanced greenhouse effect: Implications for extreme rainfall events. (Unpublished manuscript).

Grotch, S.L. and M.C. MacCracken, 1991: The use of general circulation models to predict regional climate change. *J. Climate*, **4**, 286-303.

Haarsma, R.J., J.F.B. Mitchell and C.A. Senior, 1992: Tropical disturbances in a GCM. *Climate Dyn.* (In press).

Hansen, A.R. and A. Sutera, 1990: Weather regimes in a general circulation model. *J. Atmos. Sci.*, **47**, 380-392.

Hansen, J.E., G. Russell, A. Lacis, I. Fung and D. Rind., 1985: Climate response times: Dependence on climate sensitivity and ocean mixing. *Science*, **229**, 857-859.

Hasselmann K., R. Sausen and E. Maier-Reimer, 1992: *On the cold-start problem in transient simulations with coupled ocean-atmosphere models.* Max Planck Inst. Meteor. Report, Hamburg. (In press).

Hibler, W.D., 1979: A dynamic-thermodynamic sea-ice model. *J. Phys. Oceanogr.*, **9**, 815-846.

Hibler, W.D., 1984: *Sensitivity of sea-ice models to ice and ocean dynamics.* WCP-77, WMO, Geneva.

Hoskins, B.J., H.H. Hsu, I.N. James, M. Masutani, P.D. Sardeshmukh and G.H. White, 1989: *Diagnostics of the global atmospheric circulation based on ECMWF analyses 1979-1989.* WCRP-27, WMO/TD - No. 326, World Met. Organiz., Geneva, 217 pp.

Houghton, D.D., R.G. Gallimore and L.M. Keller, 1991: Stability and variability in a coupled ocean-atmosphere climate model: Results of 100-year simulations. *J. Climate*, **4**, 557-577.

Hulme, M., 1992: A 1951-80 global land precipitation climatology for the evaluation of general circulation models. *Climate Dyn.*, **7**, 57-72.

Ingram, W.J., C.A. Wilson and J.F.B. Mitchell, 1989: Modelling climate change: an assessment of sea-ice and surface albedo feedbacks. *J. Geophys. Res.*, **94**, 8609-8622.

IPCC, 1990: *Climate Change, The IPCC Scientific Assessment.* J.T. Houghton, G.J. Jenkins and J.J. Ephraums (Eds.). Cambridge University Press, UK. 365pp.

Jaeger, L., 1976: *Monatskarten des Niederschlags für die ganze Erde.* Ber. Deut. Wetterdienstes, Nr. 139, Offenbach. 38pp.

Karl, T.R., W.-C. Wang, M.E. Schlesinger, R.W. Knight and D. Portman, 1990: A method of relating general circulation model simulated climate to the observed local climate. Part I: Seasonal statistics. *J. Climate*, **3**, 1053-1079.

Karl, T.R., R.R. Heim and R.G. Quayle, 1991: The greenhouse effect in central North America: If not now, when? *Science*, **251**, 1058-1061.

Katz, R.W. and B.G. Brown, 1991: Extreme events in a changing climate: Variability is more important than averages. *Climatic Change*. (In press).

Kelly, K.K., A.F. Tuck and T. Davies, 1991: Wintertime asymmetry of upper tropospheric water content between Northern and Southern Hemispheres. *Nature*, **353**, 244-247.

Kiegwin, L.D., G.A. Jones, S.J. Lehman and E.A. Boyle, 1991: Deglacial meltwater discharge, North Atlantic deep circulation and abrupt climate change. *J. Geophys. Res.*, **96**, 16,811-16,826.

Kiehl, J.T. and D.L. Williamson, 1991: Dependence of cloud amount on horizontal resolution in the National Center for Atmospheric Research Community Climate Model. *J. Geophys. Res.*, **96**, 10955-10980.

Kitoh, A., 1989: Persistent anomalies in the Northern Hemisphere of the MRI GCM-1 compared with the observation. *Pap. Met. Geophys.*, **40**, 83-101.

Kitoh, A., 1991: Interannual variations in an atmospheric GCM forced by the 1970-1989 SST. Part I: Response of the tropical atmosphere. *J. Met. Soc. Japan*, **69**, 251-269.

Kitoh, A., 1992: Simulated interannual variations of the Indo-Australian monsoons. *J. Met. Soc. Japan*, **70** (special edition on Asian Monsoon). (In press).

König, W., E. Kirk and R. Sausen, 1990: Sensitivity of an atmospheric general circulation model to interannually varying SST. *Ann. Geophys. Sci.*, **8**, 829-844.

Lal, M., U. Cubasch and B.D. Santer, 1992: Potential impacts of global warming on monsoon climate as inferred from a coupled ocean-atmosphere general circulation model. (Unpublished manuscript).

Latif, M., M. Flugel and J-S. Xu, 1991: *An investigation of short range climate predictability in the tropical Pacific*. Max Planck Inst. Meteor., Report No. 52, Hamburg.

Lau, K.M. and C. Peng, 1990: Origin of low-frequency (intra-seasonal) oscillations in the tropical atmosphere, Part III, Monsoon dynamics. *J. Atmos. Sci.*, **47**, 1443-1462.

Lau, N.-C., S.G.H. Philander and M.J. Nath, 1991: Simulation of El Niño-Southern Oscillation phenomena with a low-resolution coupled general circulation model of the global ocean and atmosphere. (Unpublished manuscript).

Legates, D.R. and C.J. Willmott, 1990a: Mean seasonal and spatial variability in global surface air temperature. *Theor. Appl. Climatology*, **41**, 11-21.

Legates, D.R. and C.J. Willmott, 1990b: Mean seasonal and spatial variability in gauge-corrected global precipitation. *Int. J. Climatology*, **10**, 111-127.

Lemke, P., W.B. Owens and W.D. Hibler, 1990: A coupled sea-ice mixed-layer pycnocline model for the Weddell Sea. *J. Geophys. Res.*, **95**, D6, 9513-9525.

Le Treut, H., Z.X. Li, M. Foaichon and R. Butel, 1992: The sensitivity of the LMD GCM to greenhouse forcing: an analysis of the atmospheric feedback effects. (Manuscript in preparation)

Levitus, S., 1982: *Climatological Atlas of the World Ocean*. NOAA Professional Paper 13, US Dept. Commerce, Washington, DC, USA. 173 pp.

Li, X., Q.-C. Zeng and G.C. Yuan, 1991: Paper presented at Workshop of the Monsoon Numerical Experimentation Group, Boulder, USA. October 1991.

Li, Z-X. and H. Le Treut, 1991: Cloud radiation feedback in a GCM and their dependence on cloud modelling assumptions. *Climate Dyn.* (In press).

Lindzen, R.S., 1990: Some coolness concerning global warming. *Bull. Amer. Met. Soc.*, **71**, 288-299.

Maier-Reimer, E. and U. Mikolajewicz, 1989: Experiments with an OGCM on the cause of the Younger-Dryas. In: *Oceanography 1988* A. Ayala-Castanares, W. Wooster and A. Yanez-Arancibia (Eds.). UNAM Press, Mexico City, pp87-100.

Manabe, S. and R.J. Stouffer, 1988: Two stable equilibria of a coupled ocean-atmosphere model. *J. Climate*, **1**, 841-866.

Manabe, S., M.J. Spelman and R.J. Stouffer, 1992: Transient responses of a coupled ocean-atmosphere model to gradual changes of atmospheric CO_2. Part II: Seasonal response. *J. Clim.*, **5**, 105-126.

Manabe, S., R.J. Stouffer, M.J. Spelman and K. Bryan, 1991: Transient responses of a coupled ocean-atmosphere model to gradual changes of atmospheric CO_2. Part I: Annual mean response. *J. Climate*, **4**, 785-818.

Manabe S. and R.T. Wetherald, 1967: Thermal equilibrium of the atmosphere with a given distribution of relative humidity. *J. Atmos. Sci.*, **24**, 241-259.

Marotzke, J. and J. Willebrand, 1991: Multiple equilibria of the global thermohaline circulation. *J. Phys. Oceanogr.* (In press).

McAvaney, B.J., R. Colman, J.F. Fraser and R.R. Dahni, 1991: *The response of the BMRC AGCM to a doubling of CO_2.* BMRC Technical Memorandum No.3. (In press).

McGregor, S.L. and K. Walsh, 1991: *Summertime climate simulations for the Australian region using a nested model*. Proceedings Fifth Conf. Climate Variations, Amer. Met. Soc., Denver, USA.

Mearns, L.O., 1991: *Changes in climate variability with global warming and its possible impacts*. Proceedings of the First Nordic Interdisciplinary Conference on the Greenhouse Effect, 16-19 September 1991, Copenhagen. (In press).

Mearns, L.O., S.H. Schneider, S.L. Thompson and L.R. McDaniel, 1990: Analysis of climate variability in general circulation models: Comparison with observations and changes in variability in $2\times CO_2$ experiments. *J. Geophys. Res.*, **95**, 20469-20490.

Meehl, G.A., 1990: Seasonal cycle forcing of El Niño-Southern Oscillation in a global, coupled ocean-atmosphere GCM. *J. Climate*, **3**, 72-98.

Meehl, G.A. and W.M. Washington, 1990: CO_2 climate sensitivity and snow-sea-ice albedo parametrization in an atmospheric GCM coupled to a mixed-layer ocean model. *Climatic Change*, **16**, 283-306.

Meehl, G.A., W.M. Washington and T.R. Karl, 1991a: Low-frequency variability and CO_2 transient climate change. Part I. Time-averaged differences. (Unpublished manuscript)

Meehl, G.A., G.W. Branstator and W.M. Washington, 1991b: El Niño-Southern Oscillation and CO_2 climate change. (Unpublished manuscript)

Michaud, R. and C.A. Lin, 1991: Monthly summaries of merchant ship surface marine observations and implications for climate variability studies. *Climate Dyn.* (In press)

Mikolajewicz, U. and E. Maier-Reimer, 1990: Internal secular variability in an ocean general circulation model. *Climate Dyn.*, **4**, 145-156.

Miller, M.J., A.C.M. Beljaars and T.N. Palmer, 1991: The sensitivity of the ECMWF model to the parametrization of evaporation from the tropical oceans. *J. Climate.* (In press)

Mitchell, J.F.B., 1989: The "greenhouse effect" and climate change. *Rev. Geophys.*, **27**, 115-139.

Mitchell, J.F.B., 1991: No limit to global warming? *Nature*, **353**, 219-220.

Mitchell, J.F.B. and W.J. Ingram, 1992: On CO_2 and climate: Mechanisms of changes in cloud. *J Climate.*, **5**, 5-21.

Miyakoda, K. and J. Sirutis, 1990: Subgrid scale physics in 1-month forecasts. Part II: Systematic error and blocking forecasts. *Mon. Wea. Rev.*, **118**, 1065-1081.

Mullan, A.B. and J.A. Renwick, 1990: *Climate change in the New Zealand region inferred from general circulation models.* New Zealand Meteorological Service, 142 pp.

Murphy, J.M., 1990: Prediction of the transient response of climate to a gradual increase in CO_2 using a coupled ocean/atmosphere model with flux correction. In: *Research Activities in Atmospheric and Oceanic Modelling.* CAS/JSC Working Group on Numerical Experimentation, Report No14, WMO/TD No 396, 9.7-9.8.

Nagai, T., T. Tokioka, M. Endoh and Y. Kitamura, 1991: El Niño-Southern Oscillation simulated in an MRI atmosphere-ocean coupled general circulation model. (Unpublished manuscript)

Neelin, J.D., M. Latif, M.A.F. Allaart, M.A. Cane, U. Cubasch, W.L. Gates, P.R. Gent, M. Ghil, C. Gordon, N.-C. Lau, C.R. Mechoso, G.A. Meehl, J.M. Oberhuber, S.G.H. Philander, P.S. Schopf, K.R. Sperber, A. Sterl, T. Tokioka, J. Tribbia and S.E. Zebiak, 1992: Tropical air-sea interaction in general circulation models. *Climate Dyn.* (In press).

Oberhuber, J.M., 1988: *An atlas based on the COADS data set: The budgets of heat, buoyancy and turbulent kinetic energy at the surface of the global ocean.* Max Planck Inst. Meteor., Report No. 15, Hamburg, 238 pp.

Oglesby, R.J. and B. Saltzman, 1990: Sensitivity of the equilibrium surface temperature of a GCM to systematic changes in atmospheric carbon dioxide. *Geophy. Res. Letters*, **17**, 1089-1092.

Oort, A.H., 1983: *Global Atmospheric Circulation Statistics 1958-1973.* NOAA Professional Paper 14, U.S. Dept. Commerce, Washington, DC, USA. 180 pp.

Owens, W.B. and P. Lemke, 1990: Sensitivity studies with a sea-ice mixed-layer pycnocline model in the Weddell Sea. *J. Geophys. Res.*, **95**, C6, 9527-9538.

Palmer, T.N., C. Brankovic, P. Viterbo and M.J. Miller, 1991: Modelling interannual variations of summer monsoons. *J. Clim.* (In press).

Park, C.K., D.M. Straus and K.M. Lau, 1990: An evaluation of the structure of tropical intra-seasonal oscillations in three general circulation models. *J. Met. Soc. Japan*, **68**, 403-417.

Philander, S.G.H., R.C. Pacanowski, N.-C. Lau and M.J. Nath, 1991: A simulation of the Southern Oscillation with a global atmospheric GCM coupled to a high-resolution tropical Pacific ocean GCM. *J. Clim.* (In press).

Pierrehumbert, R.T., 1991: *The mixing paradigm: Towards a Langrangian view of the general circulation.* Proceedings of First DEMETRA Meeting on Global Change, Chianciano Terme, Italy. (In press).

Pittock, A.B., A.M. Fowler and P.H. Whetton, 1991: *Probable changes in rainfall regimes due to the enhanced greenhouse effect.* Proceedings of Intn'l. Hydrology and Water Resources Symp., Perth, Australia, October 1991.

Portman, D.A., W.-C. Wang and T.R. Karl, 1990: *A comparison of general circulation model and observed regional climates: Daily and seasonal variability.* Proceedings of 14th Climate Diagnostics Workshop, NOAA, Washington, D.C., USA. pp 282-288.

Ramanathan, V. and W. Collins, 1991: Thermodynamic regulation of ocean warming by cirrus clouds deduced from the 1987 El Niño. *Nature*, **351**, 27-32.

Randall, D.A., Harshvardhan and D.A. Dazlich, 1991a: Diurnal variability of the hydrologic cycle in a general circulation model. *J. Atmos. Sci.*, **48**, 40-61.

Randall, D.A., R.D. Cess, J.P. Blanchet, G.J. Boer, D.A. Dazlich, A.D. Del Genio, M.Deque, V. Dymnikov, V. Galin, S.J. Ghan, A.A. Lacis, H. Le Treut, Z.-X. Li, X.-Z. Liang, B.J. McAvaney, V.P. Meleshko, J.F.B. Mitchell, J.-J. Morcrette, G.L. Potter, L. Rikus, E. Roeckner, J.F. Royer, U. Schlese, D.A. Sheinin, J. Slingo, A.P. Sokolov, K.E. Taylor, W.M. Washington, R.T. Wetherald, I. Yagai and M.-H. Zhang, 1991b: Intercomparison and interpretation of surface energy fluxes in atmospheric general circulation models. *J. Geophys. Res.* (In press).

Raper, S.B.C., T.M.L. Wigley and P.D. Jones, 1991: The effect of man-made sulphate aerosols on empirical estimates of the climate sensitivity. (Unpublished manuscript).

Rasch, P.J. and D.L. Williamson, 1990: Computational aspects of moisture transport in global models of the atmosphere. *Quart. J. Roy. Met. Soc.*, **116**, 1071-1090.

Rasch, P.J. and D.L. Williamson, 1991: The sensitivity of a general circulation model climate to the moisture transport formulation. *J. Geophys. Res.*, **96**, 13123-13137.

Raval, A. and V. Ramanathan, 1989: Observational determination of the greenhouse effect. *Nature*, **342**, 758-761.

Rind, D., 1991: Climate variability and climate change. In: *Greenhouse-Gas-Induced Climatic Change: A Critical Appraisal of Simulations and Observations*, Elsevier, Amsterdam, pp. 69-78.

Rind, D., E.-W. Chiou, W. Chu, J. Larsen, S. Oltmans, J. Lerner, M.P. McCormick and L. McMaster, 1991: Positive water vapor feedback in climate models confirmed by satellite data. *Nature*, **349**, 500-503.

Robock, A., R.P. Turco, M.A. Harwell, T.P. Ackerman, R. Andressen, H.-S. Chang and M.V.K., Sivakumar, 1991: Use of general circulation model output in the creation of climate change scenarios for impact analysis. (Unpublished manuscript).

Santer, B.D., U. Cubasch, K. Hasslemann, W. Bruggemann, H. Hoeck, E. Maier-Reimer and U. Mikolajewicz, 1991: Selecting components of a greenhouse gas fingerprint. In: *Proceedings of First Demetra meeting on Climate Variability and Global Change*, Chianciano Terme, Italy. 28 Oct - 3 Nov.

Schlesinger, M.E., X. Jiang and R.J. Charlson, 1991: *Implication of anthropogenic atmospheric sulphate for the sensitivity of the climate system.* J.C. Allred.and A.S. Nichols (Eds.). Proceedings of Conf.Global Climate Change, Los Alamos, New Mexico, 21-24 October.

Semtner, A.J., Jr. and R.M. Chervin, 1988: A simulation of the global ocean circulation with resolved eddies. *J. Geophys. Res.*, **93**, 15502-15522.

Semtner, A.J., Jr. and R.M. Chervin, 1991: Ocean general circulation from global eddy-resolving model. (Unpublished manuscript).

Senior, C.A. and J.F.B. Mitchell, 1992a: CO_2 and climate: The impact of cloud parametrization. (Unpublished manuscript).

Senior, C.A. and J F B Mitchell, 1992b: The dependence of climate sensitivity on the horizontal resolution of a GCM. (Unpublished manuscript).

Shine, K.P. and A. Sinha, 1991: Sensitivity of the Earth's climate to height dependent changes in the water vapour mixing ratio. *Nature*, **354**, 382-384.

Somerville, R.C.J. and L.A. Remer, 1984: Cloud optical thickness feedbacks in the CO_2 problem. *J. Geophys. Res.*, **89**, 9668-9672.

Stocker, T.F. and D.G. Wright, 1991: Rapid transitions of the ocean's deep circulation induced by changes in surface water fluxes. *Nature*, **351**, 729-732.

Stouffer, R.J., S. Manabe and K. Bryan, 1989: Interhemispheric asymmetry in climate response to a gradual increase of atmospheric carbon dioxide. *Nature*, **342**, 660-662.

Tibaldi, S. and F. Molteni, 1990: On the operational predictability of blocking. *Tellus*, **42A**, 343-365.

Tibaldi, S., 1992: Low-frequency variability and blocking as diagnostic tools for global climate models. In Proceedings of the NATO Workshop on Prediction of Interannual Climate Variations. 22-26 July, 1991, Trieste, Italy. J. Shukla (Ed.). (Also appears in Proceedings of the Fondazione S. Paolo di Torino Workshop on Oceans, Climate and Man. 15-17 April, 1991, Turin, Italy.)

Toggweiler, J.R., K. Dixon and K. Bryan, 1989: Simulations of radiocarbon in a coarse-resolution world ocean model. I. Steady-state prebomb distributions. *J. Geophys. Res.*, **94**, 8217-8242.

Trenberth, K.E. and J.G. Olson, 1988: *ECMWF global analyses 1979-1986: Circulation statistics and data evaluation.* NCAR Tech. Note TN -300 +STR, 94pp.

Tselioudis, G., W. Rossow and D. Rind, 1991: Global patterns of cloud optical thickness variation with temperature. (Unpublished manuscript).

Vinnikov, K. Ya. and I.B. Yeserkepova, 1991: Soil moisture: Empirical data and model results. *J. Climate*, **4**, 66-79.

Von Storch, H., E. Zorita and U. Cubasch, 1991: *Downscaling of global climate estimates to regional scales: An application to Iberian rainfall in wintertime.* Max Planck Inst. Meteor. Report No. 64, Hamburg.

Wang, H.J., Q.C. Zeng and X-H. Zhang, 1991a: Simulation of climate change due to doubled atmospheric CO_2 using the IAP GCM with a mixed layer ocean. (Unpublished manuscript).

Wang, W.-C., M.P. Dudek, X-Z. Liang and J.T. Kiehl, 1991b: Inadequacy of effective CO_2 as a proxy in simulating the greenhouse effect of other radiatively active gases. *Nature*, **350**, 573-577.

Wang, W.-C., M.P. Dudek and X-Z. Liang, 1992: A general circulation model study of the climatic effect of atmospheric trace gases. (Unpublished manuscript).

Washington, W.M. and G.A. Meehl, 1989: Climate sensitivity due to increased CO_2: Experiments with a coupled atmosphere and ocean general circulation model. *Climate Dyn.*, **4**, 1-38.

Washington, W.M. and G.A. Meehl, 1991: Characteristics of coupled atmosphere-ocean CO_2 sensitivity experiments with different ocean formulations. In: *Greenhouse-Gas-Induced Climatic Change: A Critical Appraisal of Simulations and Observations.* Elsevier, Amsterdam, pp. 79-110.

Washington, W.M. and G.A. Meehl, 1992: Greenhouse sensitivity experiments with penetrative cumulus convection and tropical cirrus effects. (Unpublished manuscript).

WCRP, 1988a: *Validation of Satellite Precipitation Measurements for the Global Precipitation Climatology Project* (Report of Intnl. Workshop, Washington, DC, Nov. 1986). World Climate Research Programme, WCRP-1, WMO/TD - No. 203, WMO, Geneva, 27pp.

WCRP, 1988b: *World Ocean Circulation Experiment Implementation Plan. Vol. II, Scientific Background.* World Climate Research Programme, WCRP-12, WMO/TD - No. 243, WMO, Geneva, 132pp.

WCRP, 1990a: *Report of the Fourth Session of the TOGA Numerical Experimentation Group* (Palisades, NY, USA, June 1990). World Climate Research Programme, WCRP-50, WMO/TD - No. 393, WMO, Geneva, 24pp.

WCRP, 1990b: *Report of the Second Session of the Monsoon Numerical Experimentation Group* (Kona, HI, July 1990). World Climate Research Programme, WCRP-49, WMO/TD - No. 392, WMO, Geneva, 16pp.

WCRP, 1990c: *Scientific Plan for the Global Energy and Water Cycle Experiment.* World Climate Research Programme, WCRP-40 WMO/TD - No. 376, WMO, Geneva, 83pp.

WCRP, 1991: *The Global Climate Observing System* (Report of meeting, Winchester, UK, January 1991). World Climate Research Programme, WCRP-56, WMO/TD - No. 412, WMO, Geneva, 15pp.

WCRP, 1992: *The transient climate response to increasing greenhouse gases as simulated by four coupled atmosphere-ocean GCMs.* Report of the Steering Group on Global Climate Modelling (SGGCM), World Climate Research Programme, Geneva. (Unpublished manuscript).

Weaver, A.J. and E.S. Sarachik, 1991: The role of mixed boundary conditions in numerical models of the ocean's climate. *J. Phys. Oceanogr.* (Special Edition on Modelling the Large-scale Circulation of the Ocean). (In press).

Weaver, A.J., E.S. Sarachik and J. Marotzke, 1991: Internal low-frequency variability of the ocean's thermohaline circulation. *Nature.* (In press).

Webb, D.J., P.D. Killworth, A.C. Coward and S.R. Thompson, 1991: *The FRAM Atlas of the Southern Ocean.* The Natural Environment Research Council, Swindon, UK.

Wigley, T.M.L., P.D. Jones, K.R. Briffa and G. Smith, 1990: Obtaining sub-grid-scale information from coarse-resolution general circulation model output. *J. Geophys. Res.*, **95**, 1943-1953.

Wigley, T.M.L. and S.C.B. Raper, 1990: Detection of the enhanced greenhouse effect on climate. In: *Climate Change: Science, Impacts and Policy.* Proceedings Second World Climate Conference. Eds: J. Jäger and H.L. Ferguson. Cambridge Univ. Press, UK. pp. 231-242.

Wright, P.B., 1988: *An atlas based on the COADS data set: Fields of mean wind, cloudiness and humidity at the surface of the global ocean*. Max Planck Inst. Meteor. Report No. 14, Hamburg. 70 pp.

Zeng, Q.-C., G.C. Yuan, W.-Q. Wang and X.-H. Zhang, 1990: Experiments in numerical extra-seasonal prediction of climate anomalies. *J. Chinese Acad. Sci.*, **14**, 10-15.

C

Observed Climate Variability and Change

C.K. FOLLAND, T.R. KARL, N. NICHOLLS, B.S. NYENZI, D.E. PARKER,
K.Ya. VINNIKOV

Contributors:
J.K. Angell; R.C. Balling; R.G. Barry; M. Chelliah; L. Chen; J.R. Christy;
G.R. Demarée; H.F. Diaz; Y. Ding; W.P. Elliott; H. Flohn; E. Friis-Christensen;
C. Fu; P.Ya. Groisman; W. Haeberli; J.E. Hansen; P.D. Jones; D.J. Karoly;
K. Labitzke; K. Lassen; P.J. Michaels; A.H. Oort; R.W. Reynolds; A. Robock;
C.F. Ropelewski; M.J. Salinger; R.W. Spencer; A.E. Strong; K.E. Trenberth;
S. Tudhope; S.-W. Wang; M.N. Ward; T.M.L. Wigley; H. Wilson; F.B. Wood

CONTENTS

EXECUTIVE SUMMARY

1. Globally-averaged land and ocean surface temperatures for 1990 and 1991 have been similar to those of the warmest years of the 1980s and continue to be warm relative to the rest of the record. Trends show, however, regional and seasonal diversity and not every region shows warming.

2. Continuing research into the nineteenth century ocean temperature record has not significantly altered our calculation of surface temperature warming of 0.45±0.15°C since the late nineteenth century.

3. In those land areas for which we have adequate data on maximum and minimum temperatures (approximately 25% of the global land mass), the observed warming over the past several decades is primarily due to an increase of the daily minimum (night-time) temperatures with little contribution from the daily maximum (daytime) temperatures. The source of the greater warming at night relative to that by day is not clear but could be related to enhanced cloudiness, increasing concentrations of man-made sulphur-based aerosols, increasing concentrations of greenhouse gases and possibly to residual urbanization effects in the record.

4. A new analysis of radiosonde data confirms that mid-tropospheric warming has occurred over the past several decades. Combining information from the two available analyses, the radiosonde data show a mid-tropospheric warming at the rate of 0.21°C/decade in the Northern Hemisphere and 0.23°C/decade in the Southern Hemisphere over the period 1964-1991. By contrast, the new analysis of upper tropospheric temperature changes shows less cooling than estimated in the 1990 Scientific Assessment. However, time-varying biases known to exist in radiosonde temperature instruments have yet to be quantified.

5. Microwave Sounding Unit (MSU) data provide a more complete satellite-based global dataset for tropospheric and stratospheric mean temperatures, but the record is still too short for a meaningful assessment of trends. MSU data show less warming in the mid-troposphere than do the radiosonde data since 1979, though a full analysis of inhomogeneities in the MSU data that might affect trends has not been done. Because of volcanic eruptions, the MSU data show substantial 1-2 year time-scale stratospheric (100-50 hPa) warmings exceeding 1°C on a global average. However, after combining information from the two available analyses, the longer global radiosonde record shows a lower stratospheric cooling at the rate of -0.45°C/decade over the period 1964-1991.

6. The recent eruption of Mount Pinatubo injected two to three times as much sulphur dioxide (SO_2) into the stratosphere as did the El Chichon eruption. The tropical stratosphere warmed by several degrees in response but is now cooling again. A global surface and tropospheric cooling of several tenths of a degree is possible over the next year or two due to this eruption, the amount depending on the counterbalancing warming influence of the current El Niño event in the tropical Pacific and other natural influences. However, if any cooling occurs, it will be short-lived compared with time-scales of greenhouse-gas-induced climate change.

7. Precipitation variations of practical significance have been documented in a number of regions on many time and space scales. Owing to data coverage and inhomogeneity problems, however, we cannot yet say anything new about global-scale changes.

8. Evidence continues to support an increase in water vapour in the tropical lower troposphere since the mid-1970s, though the magnitude is uncertain due to data deficiencies. An increase is consistent with the observed increase in lower tropospheric temperature. However, we cannot say whether the changes are larger than natural variability.

9. Northern Hemisphere snow cover continues its tendency to be less extensive than that observed during the 1970s when reliable satellite observations began. Its reduction relates well to simultaneous increases of extratropical Northern Hemisphere land air temperature.

10. No systematic change can be identified in global or hemispheric sea-ice cover since 1973 when satellite measurements began.

11. Some influence of solar changes on climate on the time-scale of several sunspot cycles is plausible but remains unproven.

12. It is still not possible to attribute with high confidence all, or even a large part of, the observed global warming to the enhanced greenhouse effect. On the other hand, it is not possible to refute

the claim that greenhouse-gas-induced climate change **has** contributed substantially to the observed warming. The findings that increasing concentrations of man-made tropospheric aerosols have tended to cool the climate and that decreased lower stratospheric ozone is also likely to have a cooling effect in the troposphere, help to bring the observed warming into better accord with model estimates of the warming effect of increasing greenhouse gases.

C1 Introduction

We present a supplement to Section 7 (Folland *et al.*, 1990b - hereafter referred to as S7) "Observed Climate Variations and Change" of the 1990 IPCC Scientific Assessment (IPCC, 1990). It should be read in conjunction with S7 to obtain a fuller discussion of observed climate variations and changes. The main purpose of the supplement is to introduce new findings and to update important time-series and maps contained in S7 with emphasis on large spatial scales and recent satellite evidence. Consistent with this approach, most of the references contained in Section C are confined to those published since 1989 or to those which are in press. Some work on the detection and attribution of climate change is also briefly reported, a subject previously contained in Section 8 (S8) of the Scientific Assessment "Detection of the Greenhouse Effect in the Observations" (Wigley and Barnett, 1990). Although mainly discussed in Section A of the present report, we also mention recent findings relevant to interpretation of the climate record in terms of "external" forcing factors like solar variations and tropospheric and volcanic aerosols. Discussion of the surface cooling effect of recent lower stratospheric ozone reductions is confined to Section A, as there is no unambiguous observational data available to confirm this.

C2 Palaeoclimate Variations and Change - Climates Mainly Before the Late Nineteenth Century

Several recent palaeoclimate studies are mentioned here for their probable importance in estimating natural low-frequency climate variability. Note that each series is limited to sampling a small area and most of the series are biased towards measuring one or two seasons. (When interpreting this Section, the reader may wish to refer forward to the discussion of the instrumental temperature record in Section C3.1).

A study of one thousand years of tree-ring data (Cook *et al.*, 1991) confirms a strong twentieth century warming of summer temperature in Tasmania following a pronounced cold period in the early 1900s, though the warming is still within the range of natural climate variability experienced over the past 1000 years. The strong twentieth century warming is consistent with New Zealand tree ring evidence (Norton *et al.*, 1989) and large glacial retreats there since 1860 (Salinger, 1990). However, no allowance has been made in these and similar studies for the possible fertilization effect of twentieth century increases in carbon dioxide (CO_2) on tree growth, neglect of which might lead to an overestimate of recent warming.

Oxygen isotope measurements from the northern Antarctic Peninsula have been interpreted as evidence of *warmer* temperatures during the nineteenth century compared with the twentieth century (Aristarain *et al.*,

1990). However, the isotope/temperature link is weak both physically and statistically (Peel, 1992), and accumulation rate changes, which are more directly related to *in situ* temperatures, point to *cooler* conditions in the nineteenth century (Jones *et al.*, 1992). Fragmentary evidence from expedition reports also points to cooler conditions during the first decade of the twentieth century (Jones, 1990).

A recent study of documentary evidence in China (Wang and Wang, 1991; Wang *et al.*, 1991) reveals that the seventeenth and nineteenth centuries were the coldest periods there in the last 500 years (Figure C1a and 1b). Finally, Briffa *et al.* (1992) have expanded the analysis presented in an earlier paper (Briffa *et al.*, 1990) where they use tree-ring data to reconstruct summer temperatures for northern Fennoscandia since AD 500. Their new analysis is designed to highlight greater than century time-scale variability which was largely removed by the analysis procedure they used previously. They find good evidence **in this region** for a "Medieval Climatic Optimum" (S7, p 202) around 870-1110, another warm period around 1360-1570, and a "Little Ice Age" (S7, p202) period around 1570-1750. Because of the pronounced multidecadal temperature fluctuations in their data, Briffa *et al.* (1990) suggest that greenhouse-gas-induced summer warming in Fennoscandia might not be detectable until after AD 2030.

A considerable number of instrumental records, mainly but not exclusively in Europe, extend back prior to the late nineteenth century, such as the De Bilt temperature record in the Netherlands, to 1706 (van Engelen and Nellestijn, 1992) and the Central England temperature record in England, to the late seventeenth century with homogenized daily values back to 1772 (Parker *et al.*, 1992). Such records could be combined with palaeoclimate data to

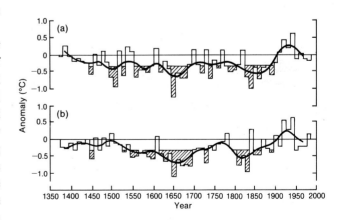

Figure C1: (a) Variations ("anomalies") of air temperature in East China (approximately 25°-35°N, 110°-122°E) since 1380, relative to 1880-1979, based on documentary evidence. The smoothed curve is a 50-year running average. Decades colder than the 1380-1879 average are shaded. (b) As (a) but for North China (35°-45°N, 110°-120°E). From Wang and Wang (1991).

provide a more detailed history of climate back to the eighteenth century (Bradley *et al.*, 1991). Thus, Lough (1991) has combined observed rainfall and river flow data with data on coastal coral growth to provide a proxy summer rainfall record for Queensland back to 1735.

C3 The Modern Instrumental Record

C3.1 Surface Temperature Variations and Change

C3.1.1 Hemispheric and Global Land Temperature

There are three independently derived, but overlapping data sets, those of Hansen and Lebedeff (1988), Jones (1988) and Vinnikov *et al.* (1990). These show noticeable differences (Elsner and Tsonis, 1991a), attributed to differences in the amounts of raw data (especially in the early parts of the record, for which the Jones data set is more comprehensive than the other two), from the methods used to ensure homogeneity of individual station records, and from methods used for spatial averaging. Jones *et al.* (1991) discuss differences between the data sets in detail. Nonetheless, the average intercorrelation between the global annual temperature anomalies from all three sets between 1881 and 1990 is 0.94.

The Jones hemispheric land air temperature series are

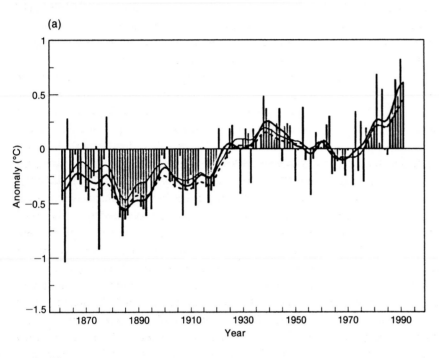

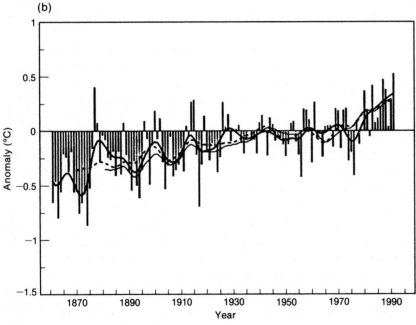

Figure C2: Land air temperature anomalies, relative to 1951-1980. Annual values from Jones (1988, updated). Smoothed curves: thick solid line = Jones (1988, updated) (1861-1991); dashed line = Hansen and Lebedeff (1988, updated) (1870-1991); thin solid line = Vinnikov *et al.* (1990, updated) (1861-1990 NH and 1881-1990 SH). (a) Northern Hemisphere; (b) Southern Hemisphere.

extended to 1991 in Figures C2a and 2b (vertical bars show annual Jones data) and the smoothed Hansen and Lebedeff and Vinnikov *et al.* series are extended to 1991 and 1990 (data from the original authors). The Jones data differ slightly from the data used in S7 because an improved method was used to convert 1951-70 anomalies to 1951-80 anomalies. Smoothed curves in this supplement, except where otherwise described, use a 21 point binomial low pass filter as mentioned in paragraph 1 on p207 of S7. In the Jones data set, 1990 was the warmest year in the Northern Hemisphere record. In the Southern Hemisphere, 1991 appears to have been the warmest, because of anomalous warmth in Antarctica, but data for some areas are incomplete. For both hemispheres, the 1980s was the warmest decade in the entire record.

Warming due to urbanization may still affect these results but is probably not serious (S7, p 209). However, a physically-based analysis of the urbanization problem is still lacking; a recently published simplified physical model of urbanization warming indicates that the problem may be more complex than hitherto thought (Oke *et al.*, 1991). Jones *et al.* (1990) have compared rural-station temperature data sets over three large regions, European parts of the Soviet Union, eastern Australia and eastern China, with widely used hemispheric data sets. When combined with earlier analyses for the contiguous United States, the regions are representative of about 20% of the land area of the Northern Hemisphere and 10% of the Southern Hemisphere and contain some of the most heavily populated areas. They indicate that urbanization influences have yielded, on average, a warming of less than 0.05°C during the twentieth century over the global land. The reasons for this result are only partly understood but, for the Jones data set, they indicate that the station-by-station quality control procedures used were fairly successful.

C3.1.2 Hemispheric and Global Sea Surface Temperature
C3.1.2.1 Ship data

In S7 two Sea Surface Temperature (SST) analyses were used: those of Bottomley *et al.* (1990), up to 1989 and those of Farmer *et al.* (1989) up to 1986, the latter now discussed in Jones *et al.* (1991). Here we show an updated time-series that uses the Jones *et al.* data and a new UK Meteorological Office analysis to 1991. This new analysis combines the Bottomley *et al.* (1990) SST data base with some Comprehensive Ocean-Atmosphere Data Set (COADS) SST values. The COADS (Woodruff *et al.*, 1987) holds substantially more surface marine observations than does the data base created by Bottomley *et al.* (1990) but both sets contain unique data (Woodruff, 1990). The COADS were used to fill missing values in the fields of monthly Bottomley *et al.* (1990) SST data, mostly in the eastern half of the Pacific. Figure C3a shows the

percentage of ocean covered by the new data and by the Bottomley *et al.* (1990) data. The new data reflect a substantial increase in coverage between the late 1870s and around 1910. Note that the new UK Meteorological Office SST data coverage exceeds that of the COADS. Improved instrumental corrections that use better models of heat transfers affecting wooden and canvas buckets have been used in the new UK Meteorological Office analysis as discussed in Folland (1991) and Folland and Parker (1992). The average magnitudes of these corrections, which vary geographically, with season and through time, are near 0.3°C (canvas buckets) and 0.1°C (wooden buckets). Although these values are small compared with the uncertainty in individual observations (Trenberth *et al.*, 1992) they are important when calculating the averages of thousands or more observations.

Another revision, that removes a little of the disagreement between Bottomley *et al.* (1990) and Farmer *et al.* (1989) data that was discussed in S7, is based on further evidence that wooden buckets may have been in predominant use at the beginning of the record. It is now assumed that 100% of buckets were wooden in 1856, the percentage linearly decreasing to 0% in 1920 (Folland and Parker, 1992). This transition from wooden to canvas buckets is nearly the same as that assumed by Farmer *et al.* (1989). However, since these authors assumed zero corrections for wooden buckets, appreciable differences between the two corrected data sets remain.

Although we feel justified in presenting a single best estimate of hemispheric and global SST changes since 1861 based on an average of the Jones *et al.* and UK Meteorological Office data (Figures C3b-d), we also show the smoothed curves for each data set. The new best estimate hemispheric and global SST curves generally lie between the continuous and dashed curves shown in Figures 7.8a and 7.8b of S7 (see note about Figure 7.8b in italics at the end of this sub-section). The addition of more Pacific data from the COADS in the late nineteenth century to the Bottomley *et al.* (1990) values has slightly increased the global average temperature in the UK Meteorological Office data set before 1900, and it remains warmer then than the Jones *et al.* data, largely owing to the positive wooden bucket corrections. Despite these "improvements", the uncertainty in the levels of nineteenth and early twentieth century SST due to data biases is 0.1°C at the least, the typical difference between the UK Meteorological Office and Jones *et al.* (1991) corrections at that time. This uncertainty increases markedly when the effects of data gaps are included.

Modern SST data may also contain non-trivial biases (Folland *et al.*, 1992), but only if these have changed significantly in recent decades would they affect **trends**. The fact that the many regional SST and night marine air temperature graphs shown in Bottomley *et al.* (1990) track

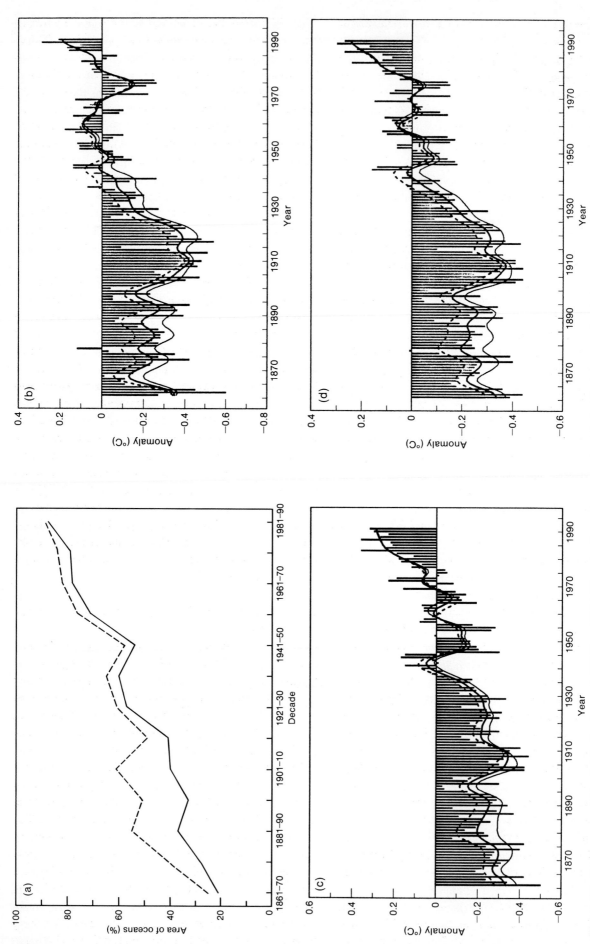

Figure C3: (a) Coverage of blended Bottomley *et al.* (1990) and COADS sea surface temperature data (dashed line) compared with that used by Bottomley *et al.* (1990) in S7 (solid line); (b) Time-series of Northern Hemisphere sea surface temperature anomalies relative to 1951-1980, based on the average of (i) an updated blend of Bottomley *et al.* (1990) and COADS, and (ii) data from Jones *et al.* (1991) (bars and heavy smoothed line). Separate smoothed series from (i) and (ii) shown by dashed and light solid lines respectively; (c) As (b) but for the Globe; (d) As (b) but for the Southern Hemisphere.

each other well in recent decades indicates that this is unlikely to be a serious problem.

Taken at face value, the increases of global mean SST between the periods 1861-1880, 1881-1900, 1901-1920, 1921-1940 and 1941-1960, and the single decade 1981-90, are now 0.43, 0.38, 0.50, 0.35 and 0.19°C respectively. This result highlights the irregularity of the warming which is largely concentrated in the periods 1920-1940 and 1975-1990 with sharp cooling between about 1900 and 1910.

Recently, Bates and Diaz (1991) have shown that even the present coverage of ship observations in the southern oceans south of 40°S is insufficient to adequately define the annual cycle of SST. Not surprisingly, therefore, SST anomaly time-series show that the Bottomley *et al.* (1990) 1951-80 climatology contains biases in parts of this area where very sparse data were blended with an earlier climatology. Though these biases probably do not much affect estimates of trends (Section C3.1.2.2), considerable uncertainty remains in Southern Hemisphere SST and SST anomaly estimates.

Due to a printing error, Figures 7.8b and 7.10b in S7 of IPCC 1990 referring to Southern Hemisphere temperatures were transposed. Thus, as printed, Figure 7.10b (p213) really shows Southern Hemisphere SST only and Figure 7.8b (p210) shows combined Southern Hemisphere SST and land air temperature.

C3.1.2.2 Reliability of the SST data

The causes and sizes of random errors in SST data have been studied by Trenberth *et al.* (1992) and Folland *et al.* (1992). They deduce that over many parts of the global ocean the signal of inter-monthly temperature variations is inadequately resolved even on large spatial scales. However, Folland *et al.* (1992) show that for seasonal averages over ocean basins, with the exception of areas south of 40°S, the resolved climatic signal is at least twice as great as the noise. For climate change and variability studies, there is an urgent need to extend these results to the annual, decadal, and century time-scales. "Frozen grid" tests show that estimates of global SST anomalies are only slightly more sensitive to changes in coverage than are the combined SST and land data. The latter (see S7, Figure 7.10d) show a surprisingly low sensitivity to the large changes in data coverage. But, as mentioned in S7, "frozen grid" tests do not adequately include the influence of those regions for which data has always been absent or very sparse, such as much of the Southern Ocean.

The SST data used in this document and in S7 are based on *in situ* observations from ships and buoys and, for the COADS, additional observations of near-surface temperatures from bathythermographs and recent data from the USA's Coastal-Marine Automated Network. During the past decade a new SST analysis has been created by blending data from satellite-borne infrared radiometers, for regions without *in situ* data (about 20% of the oceans), with *in situ* data elsewhere (Reynolds and Marsico, 1992). This blended analysis includes optimum interpolation (Gandin, 1963), takes explicit account of the presence of sea-ice and seems to successfully remove the temporally varying biases in satellite SST data relative to *in situ* data (Folland *et al.*, 1992). The identification of the satellite SST biases has resulted in a debate about the accuracy of the operational algorithms used to convert satellite radiance values to SST in the presence of atmospheric water vapour, cloudiness and episodes of global or regional contamination by volcanic or other aerosols (McClain *et al.*, 1985; Strong, 1989; Reynolds *et al.*, 1989; Bates and Diaz, 1991).

A comparison between the UK Meteorological Office *in situ* SST data and the Reynolds and Marsico data reveals a high correlation (r=0.94) between global, seasonally averaged anomalies during 1982-1990, but with a systematic difference of 0.1°C. The difference originates mostly in the Southern Hemisphere. The UK data are 0.16°C (0.03°C) warmer in the Southern (Northern) Hemisphere with a 0.85 (0.95) correlation of the seasonal anomalies. These differences originate near ice edges where they are much larger, and can be traced mainly to the fact that the Reynolds and Marsico analysis fixes SST at ice edges at -1.8°C, the average freezing point of sea water, while the UK analysis does not do this. The effects of this difference in methodology influence the analyses for some distance from the ice edges. Despite the differences, the interannual variability and the trends in the two data sets are very similar because the data bases contain much data common to both.

C3.1.2.3 Coral reef bleaching as an indicator of SST extremes

Increased reports of the bleaching of coral reefs may indicate higher SST values in many tropical regions over the last decade. The health of the coral reefs widely found through the shallow parts of tropical oceans is known to be sensitive to a number of factors, one of which is sea temperature in the top few tens of metres of the ocean (D'Elia *et al.*, 1991). Corals can tolerate a range of temperature without damage; outside this range damage is often exhibited as "bleaching". Bleaching occurs when green algae on which the coral depend are expelled from the cells of the coral when the latter are stressed. The tolerated temperature range varies with the species of coral, different species being adapted to given local conditions (D'Elia *et al.*, 1991). Thus bleaching can be a manifestation of a sea temperature that is extreme for the locality. However, local high temperatures may act with other stresses, such as pollution, to produce bleaching (Roberts, 1991). Recently many coral reef scientists have become convinced that bleaching due to elevated sea temperatures has become more common in the tropical oceans in the last decade (Brown, 1990; Glynn, 1991) though part of the increase may be due simply to better monitoring.

Several of the most severe events occurred in the tropical east Pacific and were related to the very strong 1982-1983 El Niño warming event when SST values increased several degrees above average values in this region. One hydrocoral species suffered a reduction of range, and another was probably made extinct (Glynn and de Weerdt, 1991). Since that time several other bleaching events have occurred, though not all can be related convincingly to available SST data. A more detailed comparison of coral reef bleaching and historical SST data is needed.

C3.1.3 Land and Sea Combined

Figures C4a-c show land data from the Jones analysis combined with the average of SST data from the new UK Meteorological Office analysis and the Jones et al. (1991) SST analysis for the Northern Hemisphere, Southern Hemisphere and globe respectively. The vertical bars are annual values. The method of combining the land and ocean data here is slightly different from that used in S7 (the latter is discussed in Folland, 1990). In Figures C4a-c land and ocean values have been combined more accurately and allow (approximately) for the relative areas of land and sea in every analysed grid box for every month. Figures C4a-c also show, as thinner lines, the analyses published in S7, for comparison. Values in the nineteenth century are very close to those shown in S7, and probably indistinguishably different allowing for uncertainties. The net effect of the changes is to make the long-term warming trends assessed in each hemisphere more nearly equal, with the Southern Hemisphere relatively marginally warmer in the late nineteenth century, especially around 1880, and the Northern Hemisphere unchanged. Compared to values shown in S7, larger differences in the last few years result from the addition of the warm years 1990 and 1991. S7, p212, cited increases of, or nominal linear trends in, temperature between various periods over the last century and a very recent period. A variety of warming rates occur, especially in the Northern Hemisphere. Here we note that the overall temperature increases over the globe, Northern and Southern Hemispheres between the twenty year period 1881-1900 and the latest decade 1981-90 are 0.47, 0.47 and 0.48°C respectively. This provides an estimate of the overall warming seen in the more reliable part of the instrumental record. Comparable values shown in S7 where the decade 1980-89 was compared with the twenty year period 1881-1900 were 0.45, 0.42 and 0.48°C respectively. To illustrate the small effect of including the

least reliable earlier data, the estimated changes between 1861-80 and 1981-90 are 0.48, 0.42 and 0.56°C respectively. Differences between the two sets of changes should be regarded as being due to noise in the data. The difficulty of meaningfully calculating linear trends in these data is illustrated by Demarée (1990). He shows that the Jones Northern and Southern Hemisphere land air temperature records (Figures C2a and C2b) show a statistically significant "abrupt" change in their average values around 1920, though this may also reflect a rapid change in trend evident at that time (Section C4.2.3). This step-like character of the land-based temperature record was noted earlier by Kelly *et al.* (1985). Returning to the combined data in Figure C4, the global mean warming that commenced around 1975 represents an equally sudden change in trend from about zero to a rapid warming.

Figure C4d shows the coverage of the new combined data plotted against that used in S7. The increase reflects the increase in the UK Meteorological Office SST data coverage shown in Figure C3a. Notable is the strengthened representation of the very strong El Niño warm event of 1877-78 which, being sampled over a greater area, has had a greater effect on the hemispheric and global series.

1990 and 1991 are the warmest years in the combined land/ocean temperature record, while the 1980s is the warmest calendar decade. El Niño events are known to cause warming on a global and hemispheric average (S7, p227), but there was no clear El Niño event in 1990 in contrast to a pronounced event in 1986-7 and a very strong event in 1982-3. The reasons for the warmth of 1990 must mainly be sought elsewhere. However, 1991 did contain a pronounced El Niño event. It is known that atmospheric circulation anomalies played a part in 1990; a strong westerly atmospheric circulation over the Northern Hemisphere carried unusually warm air into Northern Eurasia in early 1990. Reduced snow cover (Section C3.4.1) is also likely to have contributed to the warmth, especially in March 1990, which was by far the warmest month in the entire Northern Hemisphere land anomaly record (Parker and Jones, 1991). Note that the ranking of individual years in the surface record is not exactly the same as for tropospheric data (see Section C3.3.1).

The difference in warming rate in recent decades in the two hemispheres is discussed in the context of possible aerosol and other effects in Section A and Section C4.2. The relative temperature anomalies are quantified in Table C1. Data for 1941-50 have not been included because of poor data coverage in the Second World War (see also Figure C4d). The difference in mean decadal anomaly changed markedly between 1946-55 and 1971-80, corresponding to a relative warming of the Southern Hemisphere compared to the Northern of nearly 0.3°C between these decades. This relative warmth of the Southern Hemisphere was greatest around 1975-1980 and

A SURFACE TEMPERATURE ANOMALIES wrt 1951-80

1981-1990

< -0.75 C   -0.75 to -0.25   -0.25 to 0 C   0 to 0.25 C   0.25 to 0.5 C   0.5 to 1.00 C   > 1.00 C

(caption to plate A on page iv)   i

B SURFACE TEMPERATURE ANOMALIES wrt 1951–80
WINTER (Dec–Feb) 1981–1990

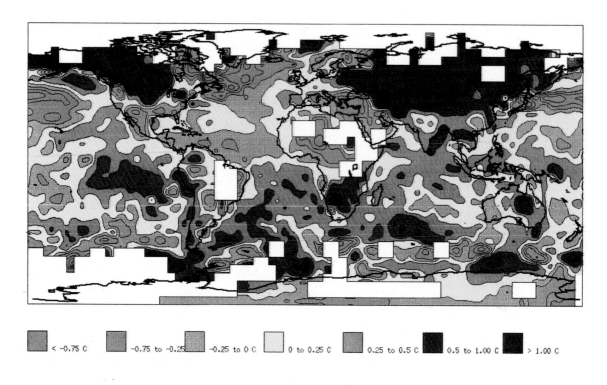

< -0.75 C	-0.75 to -0.25	-0.25 to 0 C	0 to 0.25 C	0.25 to 0.5 C	0.5 to 1.00 C	> 1.00 C

C SURFACE TEMPERATURE ANOMALIES wrt 1951–80
SPRING(Mar–May) 1981–1990

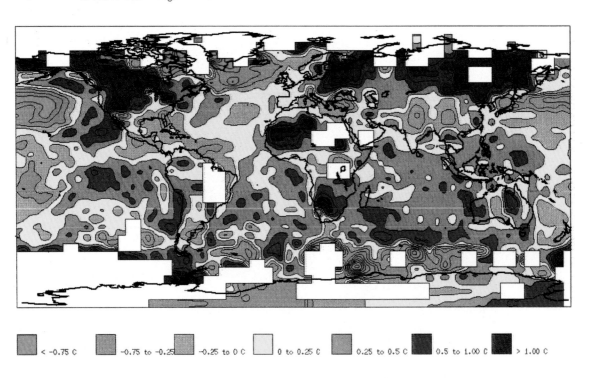

< -0.75 C	-0.75 to -0.25	-0.25 to 0 C	0 to 0.25 C	0.25 to 0.5 C	0.5 to 1.00 C	> 1.00 C

ii (captions to plates B and C on page iv)

D SURFACE TEMPERATURE ANOMALIES wrt 1951–80

SUMMER(Jun–Aug) 1981–1990

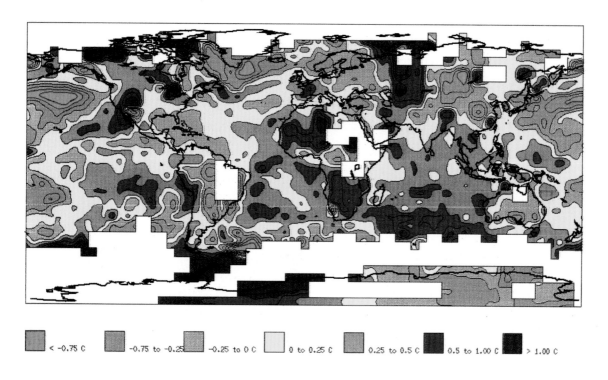

| | < -0.75 C | | -0.75 to -0.25 | | -0.25 to 0 C | | 0 to 0.25 C | | 0.25 to 0.5 C | | 0.5 to 1.00 C | | > 1.00 C |

E SURFACE TEMPERATURE ANOMALIES wrt 1951–80

AUTUMN(Sep–Nov) 1981–1990

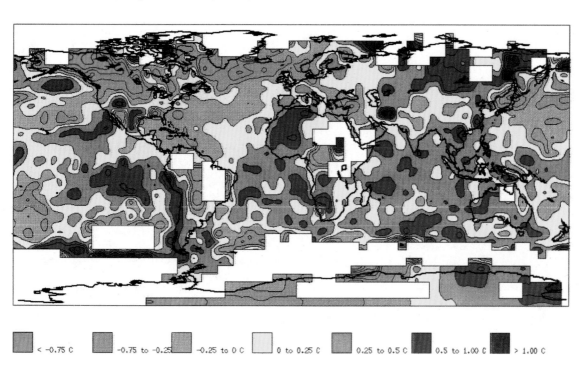

| | < -0.75 C | | -0.75 to -0.25 | | -0.25 to 0 C | | 0 to 0.25 C | | 0.25 to 0.5 C | | 0.5 to 1.00 C | | > 1.00 C |

(captions to plates D and E on page iv)   iii

CAPTIONS TO FIGURES A, B, C, D and E

Figure C5 (A): Worldwide annual surface temperature anomaly patterns, 1981–1990, relative to 1951–1980. Sea surface temperature data are an updated blend of Bottomley *et al.* (1990) and COADS. Land air temperatures provided by Jones (1988, updated). A minimum of 6 three-month seasons (Jan–Mar, etc.) with at least one month's data was required in a 5° latitude × longitude box in each half-decade, in which there had also to be at least 3 years with data; otherwise the box was treated as missing. Seasonal anomalies were averaged within each half-decade, then the two half-decadal anomalies were averaged. See legend for contour details.

Figure C5 (B): As Figure C5(A) but for Dec–Feb. A minimum of 3 seasons with at least one month's data was required in a 5° latitude × longitude box in each half-decade; otherwise the box was treated as missing.

Figure C5 (C): As Figure C5(B) but for Mar–May.

Figure C5 (D): As Figure C5(B) but for June–Aug.

Figure C5 (E): As Figure C5(B) but for Sept–Nov.

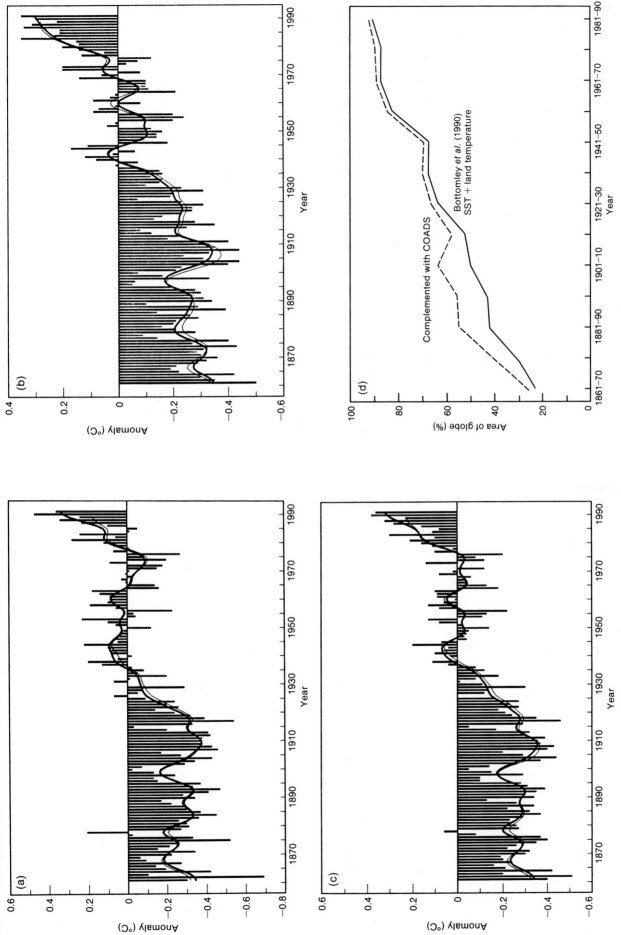

Figure C4: Combined land, air and sea surface temperature anomalies, 1861-1991, relative to 1951-1980. Land air temperatures are from Jones (1988, updated); sea surface temperatures are based on the average of (i) an updated blend of Bottomley *et al.* (1990) and COADS, and (ii) data from Jones *et al.*, 1991. Thick smoothed lines are from the updated data; thin lines are from S7: (a) Northern Hemisphere; (b) Southern Hemisphere; (c) Globe; (d) As Figure C3a but including land surface air temperature data.

Table C1: *Decadal mean surface temperature anomalies relative to 1951-80 in each hemisphere since 1946-55*

Decade	Northern Hemisphere (1)	Southern Hemisphere (2)	Difference (1)-(2)
1946-55	0.04	-0.12	0.16
1951-60	0.05	-0.06	0.11
1956-65	0.03	-0.03	0.06
1961-70	0.01	-0.03	0.04
1966-75	-0.05	0.01	-0.06
1971-80	-0.05	0.06	-0.11
1976-85	0.05	0.15	-0.10
1981-90	0.19	0.25	-0.06

the mean difference in anomalies in the last five years has returned to near zero. (The average anomalies for the three non-overlapping decades during 1951-80 are close to, but not exactly, zero due to progressive changes in data coverage.)

Considering the most recent and warmest decade, the sub-period 1986-1990 was warmer at the surface than 1981-85, though spatial patterns of temperature anomalies were, in the annual average, similar for each 5-year period. Positive anomalies in 1981-1990 were noticeably larger during December to May than in the rest of the year in both hemispheres (Figure C5). The El Niño-Southern Oscillation (ENSO) had a clear effect on the interannual variability of the global temperature, but does not explain all the observed patterns nor the fact that 1986-90 was warmer globally than 1981-1985, especially as 1986-1990 included the strong Pacific cold or "La Nina" event of 1988-1989. A cooling influence from the 1982 eruption of El Chichon may have affected the first five year period, though it is difficult to detect because of the strength of the contemporaneous 1982-83 El Niño.

C3.1.4 Worldwide Regional Temperature Anomaly Patterns

The spatial pattern of temperature anomalies at the Earth's surface for the decade 1981-1990, relative to a 1951-1980 climatology, is shown in Figure C5a. This is similar to that for 1980-1989 shown in Figure 7.13(c) of S7 and uses the UK Meteorological Office SST analysis for its ocean component. The decadal mean anomaly was assessed to be 0.22°C. Inclusion of COADS data has improved coverage over the Southern Ocean slightly. This ocean shows contrasting areas of positive and negative anomalies which may partly reflect noise in the Bottomley *et al.* (1990) 1951-80 SST climatology there. Some of the other local signals may also be affected by noise but it is not possible to quantify these problems yet (see Section C3.1.2.2 and

Trenberth *et al.* (1992)). Notable are the warmth over the middle and high latitude continents in the Northern Hemisphere, the cool anomalies over the north-west extratropical Atlantic and the predominant warmth of the Southern Hemisphere where data exist. Also shown (Figures C5b-e) are the anomalies for the constituent three-month seasons. The boreal winter map is an average from Dec 1980-Feb 1981 to Dec 1989-Feb 1990, boreal spring for Mar 1981-May 1981 to Mar 1990-May 1990, etc. Notable features are:

(1)   Dec-Feb - very large areas of positive anomalies exceeding 1°C over the high latitude Northern Hemisphere continents, with centres over 2°C, separated by notably negative anomalies over the northwestern North Atlantic, including the area around south Greenland, and also the mid-latitude North Pacific. The USA shows negative anomalies in the south, with positive anomalies near Canada. Positive anomalies over the Antarctic Peninsula are generally near 0.5°C. The global mean anomaly was 0.26°C (the most positive seasonal mean anomaly).

(2)   Mar-May - a broadly similar pattern but with a smaller area of high latitude Northern Hemisphere anomalies exceeding 1°C and positive anomalies over North Africa. Positive anomalies over the Antarctic Peninsula are stronger than in December to February. USA anomalies are positive away from the southern states, notably over the Rockies with values mostly exceeding 1°C there. The global mean anomaly was 0.24°C.

(3)   June-Aug - a broadly similar pattern to spring except that both negative and positive anomalies are seen over high latitude Asia, giving a small overall positive anomaly. Positive anomalies over the Antarctic Peninsula are at their strongest, averaging around 2°C. Anomalies over the USA are mixed but

(a)

(b)

(c)

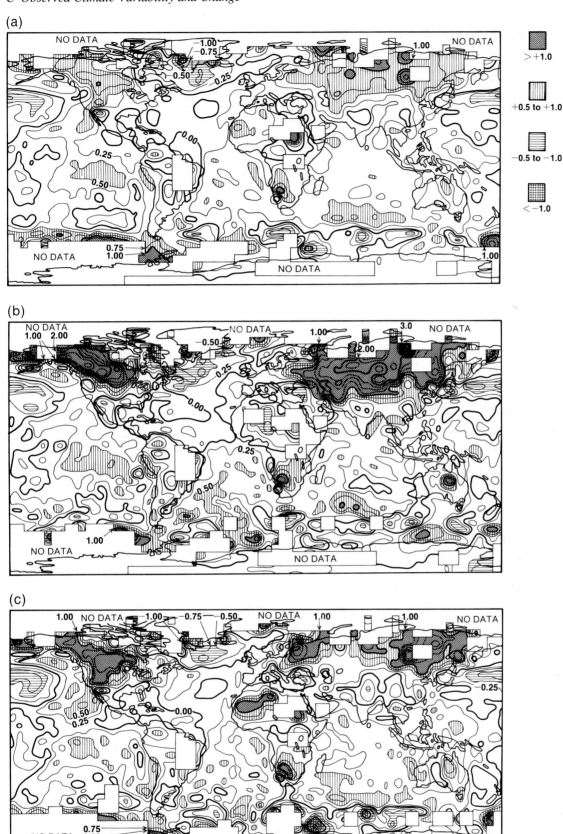

Figure C5: (a) Worldwide annual surface temperature anomaly patterns, 1981-1990, relative to 1951-1980. Sea surface temperature data are an updated blend of Bottomley *et al.* (1990) and COADS. Land air temperatures provided by Jones (1988, updated). A minimum of 6 three-month seasons (Jan-Mar, etc.) with at least one month's data was required in a 5° latitude × longitude box in each half-decade, in which there had also to be at least 3 years with data; otherwise the box was treated as missing. Seasonal anomalies were averaged within each half-decade, then the two half-decadal anomalies were averaged. Vertical hatching >0.5°C, stippling >1°C. horizontal hatching <-0.5°C, cross-hatching <-1°C. Also shown in the colour section. (Captions for (b) and (c) continued on next page.)

(d)

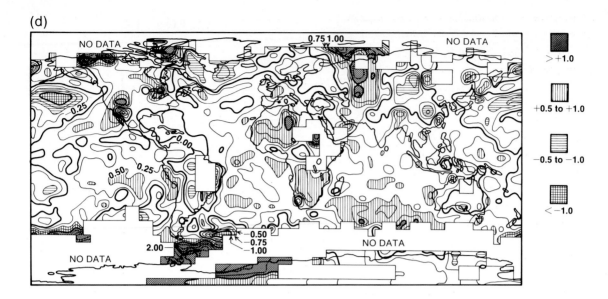

(e)

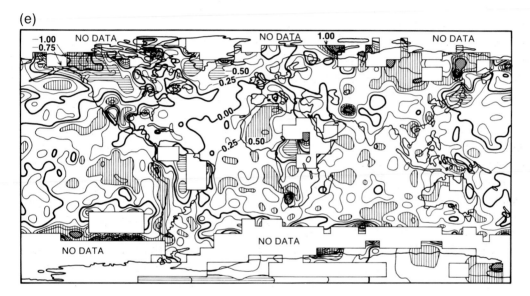

Figure C5: (b) As Figure C5a but for Dec-Feb. A minimum of 3 seasons with at least one month's data was required in a 5° latitude by longitude box in each half-decade; otherwise the box was treated as missing; (c) As Figure C5b but for Mar-May; (d) As Figure C5b but for June-Aug; (e) As Figure C5b but for Sept-Nov. Also shown in the colour section.

positive on average away from southern-most states. The global mean anomaly was 0.20°C.

(4)   Sept-Nov - major differences from summer are the negative anomalies over Alaska and most of Canada, widely cooler than -0.5°C. High latitude Asian anomalies are generally positive with a pattern rather like that of spring, but weaker. Antarctic Peninsula anomalies are weak. USA anomalies are weak and average to near zero. The global mean anomaly was 0.17°C (the least positive seasonal mean anomaly).

New land temperature series from the South Pacific and eastern Australia show that the 1980s was the warmest decade on record (Salinger and Collen, 1991; Plummer, 1991). Temperature trends in the South Pacific come from

sites where there can be little question of an urban influence. Many records are from island or remote rural sites. Land surface temperatures in eastern Australia and the South-West Pacific west of the South Pacific Convergence Zone (SPCZ) increased from the 1940s until 1990 by between 0.5° and 1.0°C. The area east of the SPCZ in the central South Pacific showed a temperature decline between 1945 and 1970, with a rapid temperature increase during the 1980s.

Temperature changes in recent decades in China have been regionally and seasonally specific (Chen *et al.*, 1991). Thus the 1980s were up to 1°C warmer than the 1951-1980 climatology in Northern China, but up to 0.5° colder in a few locations in Southern China.

Note that in Figure 7.13(c) of S7 (IPCC, 1990),

Antarctic air temperature anomalies were erroneously reduced to one tenth of their true value; this has been rectified in Figure C5. The main effect of this error was virtually to eliminate large positive anomalies (up to 1°C) over the Antarctic Peninsula. Anomalies were small elsewhere over Antarctica.

C3.1.5 Changes in the Diurnal Range of Temperature

S7 (p217) provided evidence that maximum temperatures in the USA, south-eastern Australia and China remained nearly stationary over the past several decades, but an increase of minimum temperatures was quite apparent. Karl *et al.* (1991b) have extended these time-series to include the USSR and have updated the Chinese and USA series (Figure C6). Over the last four decades, the results indicate large increases of daily minimum temperatures (0.1°C/decade) but little change of the daily maximum temperatures. Karl *et al.* (1991b) also found trends in seasonal extremes of the maximum and minimum similar to the changes of the mean maximum and minimum. Preliminary analyses for Canada and Alaska give similar results. It is unclear whether this phenomenon is global, but it is certainly a characteristic of a substantial part of the Northern Hemisphere record. However Mifsud (personal communication) finds a notable **increase** in the diurnal range in Malta over the last 60 years and Bücher and Dessens (1991), while describing a marked decrease in the diurnal range at a mountain-top station in the Pyrenees, also note that the mountain-top Sonnblick (Austria) record does not show such a change.

The results of Karl *et al.* (1991b) for the USA, the USSR, and China are appropriately area-weighted and aggregated in Table C2, which shows that the annual rate of warming is 3 to 10 times greater at night than by day, depending on the period of record chosen. Overall, summer maximum temperatures appear to have decreased in these regions.

Section 2 (S2: sub-section 2.3; Shine *et al.*, 1990) and Section A of this report describe mechanisms for increases of cloud cover, cloud albedo, and clear-sky albedo because of observed increases in sulphate aerosol. If this were occurring, there would likely be a preferential cooling effect by day, possibly leading to a reduced diurnal range in the Northern Hemisphere. This conclusion is consistent with the above results and with the finding of S7 that there was no decline in the mean daily temperature range over interior Australia and New Zealand.

Despite the known problems with changes in cloud observation practices and codes, it is likely that widely observed increases in cloud cover (S7, p230; Henderson-Sellers, 1990) have contributed to the reduced diurnal range (Plantico *et al.*, 1990; Bücher and Dessens, 1991), though not everywhere. Decreases of sunshine have been found in Germany (Weber, 1990) and the diminution of

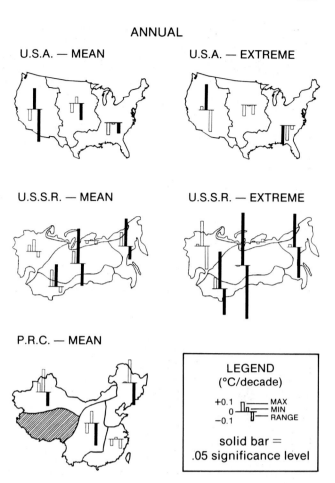

Figure C6: Trends of annual mean daily maximum and minimum temperatures and diurnal range (left) and of seasonal extreme temperatures and their differences (right). Solid bars are statistically significant at the 95% confidence level using a two tailed t-test. Data start in 1951 and finish in 1990 (USA), 1988 (China) and 1986 (USSR). There are no data available for the shaded area of southwest China.

ultraviolet radiation at low altitudes (Scotto *et al.*, 1988) coupled with enhancement at high elevation (Bruhl and Crutzen, 1989) suggests some type of increase in a lower tropospheric scattering agent.

Whatever the exact cause of the decrease in the diurnal temperature range (urbanization effects cannot be excluded and atmospheric circulation changes might also be contributing), there is an increasingly urgent need to reinterpret the global land record, at least regionally, in terms of changes in maximum (daytime) and minimum (night-time) temperatures. Care will need to be taken that artificial changes in the diurnal range are properly accounted for, such as have already been done for the USA where progressive automation of the climate observing network has taken place accompanied by a change of thermometer screen (Quayle *et al.*, 1991). The resulting non-climate related mean change in measured diurnal range over the USA is assessed to be -0.7°C (note that Karl

Table C2: *Area-weighted aggregate of temperature trends, °C/100 years (°C/40 years) for the USA, USSR and Peoples'*
Republic of China
a) Using records for 1951-90 (USA), 1951-86 (USSR), 1951-88 (China)

Season	Mean maximum (day)	Mean minimum (night)	Mean max minus min (diurnal range)
Winter	0.7 (0.3)	2.4 (1.0)	-1.7 (-0.7)
Spring	2.1 (0.8)	3.2 (1.3)	-1.2 (-0.5)
Summer	-0.7 (-0.3)	0.5 (0.2)	-1.2 (-0.5)
Autumn	0.1 (0.0)	2.2 (0.9)	-2.0 (-0.8)
Annual	0.6 (0.2)	2.0 (0.8)	-1.5 (-0.6)

b) Using records for 1901-90 (USA), 1936-86 (USSR), 1951-88 (China)

Season	Mean maximum (day)	Mean minimum (night)	Mean max minus min (diurnal range)
Winter	0.6 (0.2)	1.8 (0.7)	-1.2 (-0.5)
Spring	0.6 (0.2)	1.5 (0.6)	-0.8 (-0.3)
Summer	-0.4 (-0.2)	0.4 (0.2)	-0.8 (-0.3)
Autumn	-0.6 (-0.2)	0.7 (0.3)	-1.2 (-0.5)
Annual	0.1 (0.0)	1.1 (0.4)	-0.9 (-0.4)

et al., 1991b, analysed the corrected USA data). Meanwhile, there is a need for comparisons with results from general circulation models forced with increased greenhouse gases. Section B4.2 gives an initial discussion.

C3.2 Precipitation and Evaporation Variations and Changes

For many areas, the variability of precipitation is so large that it is virtually impossible to detect important changes until well after the occurrence of changes that have practical (e.g., agricultural or economic) significance. This is highlighted in recent work by Karl *et al.* (1991a) for the USA and Nicholls and Lavery (1992) for Australia.

C3.2.1 Precipitation Over Land

As discussed in S7, p220, raingauges have tended to underestimate precipitation, particularly snowfall (solid precipitation). Legates and Willmott (1990) and Soviet researchers (World Water Balance, 1978) have estimated that global precipitation over land is on average underestimated by 10% to 15%. Progressive improvements to instrumentation have tried to remedy this and have introduced artificial, systematic, increases in precipitation. Recent efforts to automate measurements may again be reversing this trend. Thus long-term variations and trends should be interpreted cautiously.

A selection of regional time-series was given in S7

(p219). These have been updated, but no new conclusions can be drawn. For example, precipitation over the Sahel region of North Africa (Figure 7.16b of S7) has remained well below the long-term average in 1990 and 1991 (Rowell *et al.*, 1992), extending the drought epoch there into its third decade. Changes in rainfall in the Sahel have been much larger than can be accounted for by instrumental problems. The Sahel has experienced the largest observed regional percentage change in precipitation between the two thirty year periods 1931-60 and 1961-1990: a decline of 30% (Hulme *et al.*, 1992). The latter period includes the relatively moist 1960s (S7, Figure 7.16b and Demarée and Nicolis, 1990), so the "real" change in average since 1931-1960 may be larger. The only other major region known to show a notable long-term trend in precipitation is the USSR (Figure 7.16a of S7). Area-averaged precipitation over the same region (37°-70°N, 25°-140°E) has recently been revised to eliminate minor spurious trends and updated to 1990 (Groisman *et al.*, 1991). Overall, the effects of these corrections are small and the conclusion in S7 that there has been a notable increase of precipitation over the USSR south of 70°N during the last century is unaffected.

Although a number of regional rainfall fluctuations on decadal time-scales have been found, some of practical significance, there is, as yet, no firm new evidence of global-scale multidecadal rainfall trends.

C3.2.2 *Precipitation Over the Oceans*

In S7, p220, a discussion was presented concerning the likelihood that precipitation had increased over the global tropical oceans since 1974, based on an analysis of satellite outgoing longwave radiation (OLR) data by Nitta and Yamada (1989), with an opposing opinion expressed by Arkin and Chelliah (1990). Chelliah and Arkin (1992) have now shown that much of the decreasing trend in OLR between 1974 and 1991, which appears to indicate an increase of tropical oceanic rainfall, can be unambiguously related to satellite instrumental factors rather than a real OLR increase. Therefore an increase in rainfall over the tropical oceans since 1974 remains unproven.

C3.2.3 *Evaporation from the Ocean Surface*

Estimates of trends in evaporation over the ocean are unlikely to be reliable until constant biases (Isemer and Hasse, 1991) and time-varying biases in wind speed measurement are properly accounted for (Cardone *et al.*, 1990; Ward, 1992; S7, Section 7.5.3, p220), and a proper treatment of ocean temperature at the surface interface (the ocean "skin") is included. In addition, the uncertainties in SST, near-surface air temperature and humidity data must all be allowed for, especially as estimates of evaporation changes made so far include ship-measured daytime air temperature data which may have considerable absolute and some time-varying biases as discussed in S7, p211. Furthermore, inadequate ventilation of some thermometer screens on ships, and the effect of the ship itself, can affect calculations of specific humidity. Despite these data problems, two studies have recently been made of latent heat flux trends, with some allowances for artificial trends in wind speed data. Because an increase in evaporation associated with warming is potentially a very important climatic feedback, we discuss these studies despite their drawbacks.

Following initial calculations that appeared to show an increase of the evaporation rate from parts of the tropical oceans between 1949 and 1979 (Flohn *et al.*, 1990a, b), a revised analysis has been carried out to investigate the sensitivity of the results to varying assumptions about the reality of increasing near-surface wind speed over the oceans (Flohn *et al.*, 1992). Figure C7a shows calculations of the evaporation trend in the tropical oceans between 10°S and 14°N, expressed as a linear rate between 1949 and 1989. Only the main shipping lanes have been studied. If it is assumed that the observed near-surface wind speed trends should be reduced to 50% of their measured values over the areas sampled, an average increase of about 18Wm⁻² of heat input into the atmosphere is implied over the belt as a whole. Calculations for sub-sections of the zone (warmest oceans, upwelling regions and Atlantic sector) give similar results.

In a further study, Fu and Diaz (1992) show an apparent upward trend of about 1% per year in integrated mean

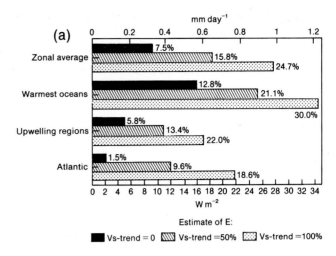

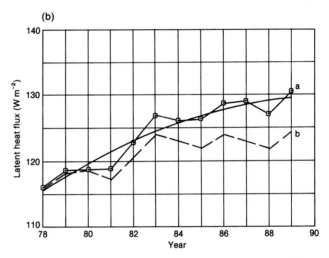

Figure C7: (a) Changes in evaporation rate, E, from the tropical oceans (10°S-14°N) between 1949 and 1989, based on COADS data after multiplying the wind speed trend, Vs, by zero, 0.5 and 1.0. Taken from Flohn *et al.* (1992): (i) whole zone; (ii) warmest oceans (66°E-160°E); (iii) upwelling regions, and (iv) Atlantic sector; (b) Trends in global annual oceanic latent heat flux (Wm⁻²), 62°N-42°S from Fu and Diaz (1992). Curve a (and smoothed curve): using unadjusted winds; Curve b: using winds reduced following Cardone *et al.* (1990).

oceanic latent heat flux during the period 1978-1989. This increase (curve a in Figure C7b) is due primarily to an overall increase in reported wind speed over the global oceans of about 0.5 ms⁻¹. Allowing for an artificial component in the increase of wind speed (Cardone *et al.*, 1990), it is estimated that the observed wind speed trend should be reduced to between 65% and 50% of its apparent value, giving the reduced trend in latent heat flux shown in curve b of Figure C7b. This corresponds to a global increase in latent heat flux into the atmosphere between 1978 and 1988 near 5Wm⁻²/decade. Similar results are obtained by Flohn *et al.* (1992) over part of the tropical North Atlantic. Although there is no indication of bias in the surface pressure measurements, possible hidden trends in pressure gradients in Fu and Quan's analyses due to variations in data sampling need investigation. Overall, it is uncertain whether the apparent enhancement of evaporation from the global oceans shown in Figures C7a and b is a real climatic signal that has accompanied the recent global oceanic warming seen in Figures C3 and C5.

An increase of tropical precipitation might be expected to accompany increases in evaporation. However, the results of Chelliah and Arkin (1992) discussed in Section C3.2.2 do not confirm this idea. So, although an increase in the hydrological cycle in the last 10-20 years is plausible, it is not proven. A rigorous analysis of the physical consistency of the local values of, and trends in, the various data used to calculate evaporation is essential for further progress.

C3.3 Tropospheric and Lower Stratospheric Variations and Change

C3.3.1 Temperature

C3.3.1.1 Radiosonde data

There are now two independently-derived data sets, that of Angell (1988) used in S7 (p220-222) and a new compilation due to Oort and Liu (OL) (1992). Both data sets may suffer from data-quality problems, because adequate studies of possible time-varying biases in radiosonde temperature data have not been made (Elliott and Gaffen, 1991). The annual data derived from the Angell analysis uses a different definition of the calendar year from all other analyses in Section C, being based on December-November. This will slightly reduce correlations between annual values of the Angell and other data sets. Unless otherwise stated, the term "lower stratosphere" used in the remainder of Section C refers to that part of the atmosphere between the 50 and 100 hPa levels.

The spatial representativeness of the Angell radiosonde data set has been examined by Trenberth and Olson (1991) in a study of how well the widely spread but sparse 63-station network describes regional and global climatic changes. By comparing its interseasonal and interannual climatic statistics over 1979-1987 with those from a complete global data set provided by the European Centre for Medium Range Weather Forecasts, they found that correlations between the two data sets were generally quite high, but that root-mean-square errors for Angell's extratropical zones were of the same order as the interannual climatic signals being studied. Angell's data also showed systematically enhanced interseasonal and interannual variability in the extratropics because of the limited spatial sampling.

The OL data are derived from up to 800 individual site records from the global radiosonde network interpolated onto a regular latitude-longitude grid. In terms of data density, the OL data base is therefore an improvement on that of Angell.

Figures C8a-c present updated annual global series of temperature anomalies from a 1964-1989 average for the surface and 850-300 hPa layer (a), the 300-100 hPa layer (b), and the 100-50 hPa layer (c), using Figure C4 for the surface. The data sets are, unfortunately, short but are important because they have been used in initial studies of the greenhouse-gas detection problem (Section C4). Although the radiosonde coverage is adequate from 1958 in the Northern Hemisphere, it is only complete enough in the Southern Hemisphere since 1964. Global values can therefore only be estimated since 1964, though Angell used incomplete Southern Hemisphere data to extend his global series back to 1958.

Correlations (1964-1989) between the OL and Angell annual global series are high, 0.97, 0.91 and 0.90 at 850-300 hPa, 300-100 hPa and 100-50 hPa respectively. However, the absence of some extensive areas may have allowed higher correlations than would be obtained between either series and a globally-complete series. Note that surface values show substantially less interannual variability. The correlation between the annual OL 850-300 hPa series and the surface (ocean and land) series is 0.94; Angell's series show slightly greater interannual variability than that of OL for the globe and (not shown) for both hemispheres. Angell's data also tend to show greater cooling than those of OL in the 300-100 hPa layer, mainly in the Northern Hemisphere but also in the Southern Hemisphere lower stratosphere 100-50 hPa layer, but less cooling in the 100-50 hPa layer in the Northern Hemisphere. OL's trends in annual mean temperature for the globe were 0.21, -0.08 and -0.41°C/decade (1964-1989) for the three layers in ascending height order, with similar trends in each hemisphere. Angell's corresponding trends for the period 1964-1991 were 0.24, -0.16 and -0.52°C/decade. Combining values from the two analyses by using the more comprehensive OL data to 1989 and the Angell data in 1990 and 1991, our best estimate for trends in the three layers between 1964 and 1991 are Northern Hemisphere: 0.21, -0.05 and -0.38°C/ decade; Southern Hemisphere: 0.23, -0.13 and -0.53°C/ decade and Globe: 0.22, -0.09 and -0.45°C/decade. The trend in the surface data over this period was 0.16°C/decade.

The warming trend in the mid-tropospheric layer and the cooling trend in the lower stratosphere are significant in both data sets at better than the 1% level, allowing for autocorrelation of the data. The warming trend at the syrface is also significant at the 1% level. However, the cooling trend in the globally-averaged 50-100 hPa layer may be exaggerated because the data begin in 1964 which was very warm in the tropical stratosphere following the eruption of Agung in 1963 (Newell, 1970). Inspection of Figure C8 suggests that after the influence of Agung has been removed, a slow cooling trend followed until 1982 when a temporary warming occurred due to the eruption

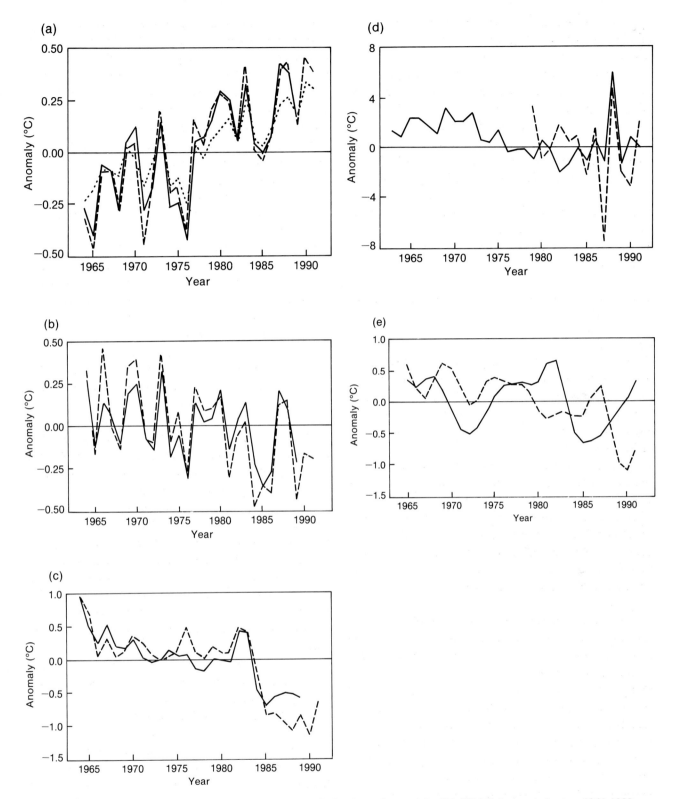

Figure C8: (a) Annual global temperature anomalies since 1964 for the surface and the 850-300 hPa layer relative to a 1964-1989 average. Surface temperatures (dotted line) are based on Figure C4. Temperatures aloft are based on the radiosonde data of Oort and Liu (1992) to 1989 (solid line) and of Angell (1988) updated to 1991 (dashed line); (b) Annual global temperature anomalies since 1964 for the 300-100 hPa layer relative to a 1964-1989 average from two analyses. Temperatures are based on the radiosonde data of Oort and Liu (1992) to 1989 (solid line) and Angell (1988) updated to 1991 (dashed line); (c) as (b) but for the 100-50 hPa layer; (d) Austral spring (Sept-Nov) temperature anomalies for the 100-50 hPa layer relative to a 1979-1991 average for the south polar cap: (i) from Oort and Liu (1992) to 1989 and from Angell thereafter for 60°-90°S (solid line); (ii) from MSU Channel 4 data for 62.5°-90°S (dashed line); (e) Smoothed annual zonal mean 30 hPa temperature anomalies relative to a 1964-1989 average for 80°N (dashed) and 20°N (solid). A low pass binomial filter with 5 terms was used to suppress variations of less than 3 years. Updated from Naujokat (1981).

of El Chichon. After this there was a sharp cooling to 1985-87. Trenberth and Olson (1989) found a lower stratospheric cooling trend in spring over the South Pole and McMurdo Sound, extending to January at 100 hPa, though there were no tropospheric trends. In fact, radiosonde data for 100-50 hPa for the south polar cap show some cooling in all seasons except autumn over the last decade or more. However, in austral spring, when the deepest ozone holes have been reported (SORG, 1991), there have been very strong interannual fluctuations of radiosonde and satellite Microwave Sounding Unit temperature (see Section C3.3.1.2) since 1985 (Figure C8d). Although, as a result, the trend in Figure C8d is not significant, the general level of lower stratospheric south polar cap temperature in spring was several degrees lower in the 1980s than the 1970s.

Low-pass filtered series of 30 hPa temperature for latitude belts of the Northern Hemisphere show weak downward trends (Figure C8e, updated from Naujokat, 1981). At many latitudes there is a suggestion of an oscillation almost in phase with the cycle of solar activity whose length over this period was quite close to 11 years, seen here in the solid curve for 20°N (see also Section C4.2.1): this should be taken into account when 30 hPa temperature trends are determined (van Loon and Labitzke, 1990; Labitzke and van Loon, 1991, 1992). Warming at 30 hPa associated with the 1982 eruption of El Chichon is especially noticeable at lower latitudes (e.g., the solid curve in Figure C8e). Data for late 1991 (not shown in Figure C8e) show another pronounced warming of up to several °C at 30 hPa. The warming, centred near 20°N and confined to regions south of 45°N, was almost certainly due to the eruption of Mt. Pinatubo in June 1991.

C3.3.1.2 Satellite microwave sounder data

S7 reported early results concerning recent tropospheric temperature trends and interannual variations from a valuable new data set derived from satellite measurements of the microwave emission of radiation to space from atmospheric oxygen (Spencer and Christy, 1990; Figure 7.17d, p221). The new technique, which uses data from the Microwave Sounding Units (MSU) on the TIROS-N series of satellites (Spencer *et al.*, 1990), measures temperature over layers of the atmosphere. Channel 2 data of the MSU is weighted towards temperature over a substantial thickness of the troposphere, but is also influenced by the stratosphere and the character of the ground surface. By using sets of Channel 2 data with different earth viewing-angles it is possible to create a new data set called Channel 2R that is mostly (not completely) weighted towards levels below 350 hPa, mainly to levels between 500 and 1000 hPa, though at the cost of a slight loss in reproducibility. In addition, Channel 4 measures lower stratospheric (mainly 30-150 hPa) temperatures. The MSU instrument is inde-

pendently calibrated, so is not influenced by conventional instruments such as radiosondes, though biases, resulting from changes of satellites and equatorial crossing-times, may remain. The main advantages of the MSU system are its global coverage and a lower value of standard error at most grid points compared with radiosonde data (Spencer and Christy, 1992a,b).

A particularly important aspect of MSU data in the context of this assessment is their ability to detect temperature trends with relatively high accuracy in the layers they measure, though the MSU record is still very short (13 years). Figure C9 shows lightly smoothed monthly global Channel 4 lower stratospheric temperature anomalies relative to the complete period 1979-91. Over this short period very large, temporary but highly coherent warming effects of the volcanoes El Chichon (1982) and Mt. Pinatubo (1991) dominate the record. This makes the detection of a global trend in MSU lower stratospheric data difficult.

C3.3.1.3 Comparisons of satellite microwave, radiosonde, and surface temperature data

The correlation between 5° latitude × 5° longitude monthly time-series of surface-based temperature anomalies and of tropospheric temperature anomalies from the MSU is near zero over parts of the tropical oceans but is much higher over major land masses and oceanic areas of high variability (Trenberth *et al.*, 1992). The differences are thought to be partly due to data sampling problems in the non-MSU data sets, partly to real physical differences between surface temperatures and the mid-tropospheric temperatures, and possibly partly to uncertainties in the MSU data due to surface emissivity variations. There are also some divergences between the MSU and the radiosonde data. Consequently, the ranking of recent very warm years in the lower atmosphere and at the surface

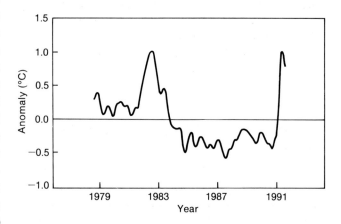

Figure C9: Smoothed monthly global MSU Channel 4 lower stratospheric temperature anomalies, for 1979 to 1991 relative to a 1979-1991 average. A binomial filter with 5 terms was used.

depends on which record is used, what level is being referred to and how much uncertainty is attached to each value.

Figure C10 compares annual global temperature anomalies for 1979-91 for the mid-troposphere from MSU Channel 2R, with values for the 850-300 hPa layer from radiosondes using Oort and Liu to 1989 and then Angell to 1991, and for the surface using the data in Figure C4c. Table C3 presents the intercorrelations (r) for MSU 2R (suffix m), radiosonde (suffix sd) and surface data (suffix sf), their linear trends (τ) in °C per decade, and the standard deviations, σ, of the annual values. Note, however, that the linear trends for 1979-1991 are not fully representative of those for the longer term.

At this stage we merely note that the calculated linear trends for 1979-1991 are different, though all but one are positive. The correlations between the surface and the MSU data are lower than those with the radiosonde data. The correlations may be reduced by the fact that the MSU

data samples regions not sampled or poorly sampled by the radiosonde and surface data, especially in the Southern Hemisphere, but further investigation is needed to clarify this. Much, though not all, of the difference in the trends between the MSU and the other data comes from disagreements in the annual anomalies for 1979, 1980 and 1981. It is clear that the data sets have some different characteristics. The short period of overlap (1979-1991), differences in the variables being measured and doubts about the year-to-year consistency of the data sets, prevent reliable assessment of the differences in trends. Reliable assessment of future trends in MSU data will require the compatibility of new MSU instrumentation with that used at present. The continued availability of all three data sets is very desirable to help reduce the problems noted above.

C3.3.2 Atmospheric Moisture

For previous discussions, refer to S7, p222 and S8, p251; in the current volume Section B3.2 gives a more detailed discussion of the likely role of water vapour during a greenhouse gas-induced warming. Water vapour is the greenhouse gas in greatest abundance and is responsible for the largest single contribution to greenhouse warming of any of the constituents of the atmosphere in the current climate.

The sensitivity of the surface temperature to height-dependent moisture changes was examined by Shine and Sinha (1991) who point out the importance of changes in water vapour content **throughout** the depth of the troposphere. They show that changes in much of the mid-troposphere are important as they tend to be large due to the relatively great amounts of water vapour there. However, despite the small absolute amount of water vapour in the upper troposphere and stratosphere, they show that changes at this level can also have a significant effect on the radiation forcing of climate. Furthermore, an increase in moisture in the stratosphere could lead to the creation of further polar stratospheric clouds which have an important effect on ozone depletion in the presence of chlorine derived from chlorofluorocarbons. Stratospheric water vapour is also important in the conversion of SO_2 to sulphuric acid droplets which can cool surface climate after a sulphur-rich volcanic eruption. Furthermore,

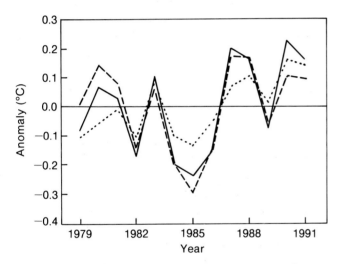

Figure C10: Comparison, for 1979-1991, of annual global temperature anomalies from (i) MSU Channel 2R for the lower troposphere (dashed line); (ii) radiosonde data for the 850-300 hPa layer from Oort and Liu (1992) to 1989 then from Angell (1988, updated) (solid line); (iii) surface data from Figure C4 (dotted line). Anomalies are referred to a 1979-1991 average in each case.

Table C3: *Annual MSU 2R, radiosonde 850-300hPa and surface temperature anomalies, 1979-1991: intercorrelations, trends and standard deviations*

	$r_{sd, m}$	$r_{sd, sf}$	$r_{m, sf}$	τ_m	τ_{sd}	τ_{sf}	σ_m	σ_{sd}	σ_{sf}
NH	0.96	0.91	0.84	0.12	0.21	0.23	0.18	0.18	0.15
SH	0.86	0.71	0.42	-0.02	0.13	0.14	0.13	0.14	0.08
Globe	0.93	0.90	0.74	0.06	0.17	0.18	0.15	0.16	0.10

oxidation of methane generates stratospheric water vapour; so increases in methane, from whatever causes, can lead to an increase in stratospheric moisture.

In an examination of strategies for detecting greenhouse gas signals, Barnett *et al.* (1991) find tropospheric moisture to be one of the most effective variables to monitor. However, monitoring moisture presents many difficulties and, since its variation is physically linked to changes of temperature, it does not provide an independent measure of greenhouse forcing. The residence time of water vapour in the atmosphere is short, about 10 days, so it is not well mixed. If the water vapour in the air were all condensed, the average depth of the condensate (the precipitable water, PW), would be about 2.5cm. Above the polar regions the mean PW is about 0.5cm, and above the equatorial regions it averages about 5cm. Half the moisture in the atmosphere lies between sea level and 850 hPa and less than 10% resides above the 500 hPa level. Therefore, observations at many places and levels are required to adequately study changes in water vapour likely to be climatically important.

Measurement problems make trends of water vapour difficult to determine. Most of the existing knowledge about tropospheric water vapour comes from routine radiosonde observations. Unfortunately, radiosonde data have been affected by changes in humidity sensors and reporting procedures, and some sensors and retrieval algorithms do not provide useful results in the low temperatures and very dry conditions of the upper troposphere and lower stratosphere. This makes it difficult to separate climatic changes from changes in the measurement programme (Elliott and Gaffen, 1991). Nevertheless, with careful attention to these problems some deductions can be made.

Recently Elliott *et al.* (1991) have documented a moisture increase in the lower troposphere over the equatorial Pacific from 1973-1986; additionally Gaffen *et al.* (1991), using more stations and an analysis of individual moisture patterns, have found an increase in moisture in the tropics. At 850 hPa, this study found an increase of specific humidity of around 10% between 1973 and 1986, though this value is very uncertain as the scatter in the data is comparable to the trend and most of the change occurred during the short interval 1977-1980. The study also detected a signal of the ENSO phenomenon (S7, p226) in the humidity data.

Thus, there is limited observational evidence suggesting an increase in lower tropospheric moisture content in tropical regions over the last two decades. These results support the apparent increase of evaporation from the oceans (Section C3.2.3). A much more comprehensive atmospheric moisture monitoring system is needed, such as may be provided by the Global Energy and Water Cycle Experiment (GEWEX) (WMO, 1991).

C3.4 Variations and Changes in the Cryosphere

This section should be read in conjunction with Section C2, where recent work on oxygen isotopes in Antarctic land-ice is discussed, and with S7, Section 7.8. Section 9.4.3 of S9 (Warrick and Oerlemans, 1990) and the brief review by Haeberli (1990) should be consulted on permafrost and other cryospheric features.

C3.4.1 Snow Cover

Snow extent is very variable; it is affected by temperature through atmospheric circulation variations which also influence the quantities of solid and liquid precipitation.

The Northern Hemisphere snow extent anomaly time-series shown in Figure 7.19 of S7 (p224) has recently been revised. A change in analysis technique had resulted in snow extents analysed prior to 1981 being slightly too large relative to later values (Robinson *et al.*, 1991). A revised series from 1973 to 1991 is given in Figure C11. This still shows a modest decrease in snow cover since the 1970s but with a reduced magnitude: mean values over 3 years dropped by just over 2 million km^2 from the mid-1970s to the end of the 1980s, about 8% of the total area. In Figure 7.19 of S7, a decline of about 3 million km^2 was indicated. The new snow extent values have good support from a parallel plot of extratropical Northern Hemispheric land air temperature (Figure C11). The correlation between the monthly (September to May only) anomalies of snow cover and temperature is -0.41, but between unsmoothed annual (average of Sept-May) anomalies it is -0.76 because presumably influences other than those of temperature partly cancel. Snow extent has been especially low in spring since 1987, most notably in spring 1990. However, the snow extent record is still much too short to distinguish a possible greenhouse signal from natural variability.

C3.4.2 Mountain Glaciers

A major result of S7 (see Executive Summary) was the conclusive evidence for a worldwide recession of mountain glaciers over the last century or more. This is among the clearest and best evidence for a change in energy balance at the Earth's surface since the end of the last century. It provides sufficient support to the various independent but far from perfect records of global temperature to show that global warming has indeed occurred over the last century (Haeberli, 1990). On a regional scale, however, the influence of climate can be much more complex, especially on decadal and lesser time-scales where precipitation fluctuations may be important or even dominant.

Direct information on mass balance is available from a small number of glaciers in the European Alps where these observations started during the late nineteenth century. The average annual loss of specific mass (mass per unit area) amounts to between 0.2 and 0.6 m water equivalent

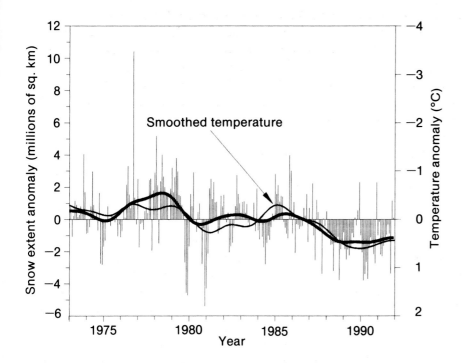

Figure C11: Northern Hemisphere snow extent anomalies relative to 1973-1991 (from NOAA, USA) (vertical bars and heavy line) and Jones (1988, updated) land air temperature anomalies relative to 1951-80 north of 30°N (thin line). Smooth lines generated from a 39-point binomial filter applied to the monthly data. Note the inverted temperature scale.

(Patzelt and Aellen, 1990). Regularly monitored glaciers, however, account for less than 1% of the total number of glaciers worldwide. Much more numerous are measurements of glacier length reductions which can be converted into mass balance values using a combination of continuity analyses and data on initial glacier length. They confirm the representativeness of the small number of direct long-term mass balance measurements (Haeberli, 1990). Aerial and ground surveys indicate that most observed mountain glaciers are distinctly smaller than a century ago (Haeberli and Müller, 1988; World Glacier Monitoring Service, 1991; Williams and Ferrigno, 1988-1991).

Some of the techniques used to estimate the overall glacier mass decrease in the Alps since the middle of the nineteenth century are controversial. Recently, Haeberli (1990) has estimated a 50% decrease in mass. Though this is subject to marked uncertainty, a relatively large decrease is likely. This drastic change is the consequence of an upward shift in equilibrium line altitude by only 100m or less. Twentieth-century melt rates of Alpine glaciers are an order of magnitude greater than average melt rates at the end of the last ice age (20,000 to 10,000 years ago).

Statistical analysis of spatial and temporal variations in mass balance series from glaciers on various continents (Reynaud *et al.*, 1984) indicates that the historical behaviour in the Alps may be representative of glaciers in general, confirming the great sensitivity of glacier volume to climate variations. A global-mean reduction in volume may be estimated from average ice volume data in S9

(Table 9.3) and the estimated sea level rise contribution from mountain glaciers and small ice-caps (Meier, 1984). This gives a reduction of 13±9% in volume over the last century. Despite this, the retreat of Alpine glaciers has not been uniform on decadal time-scales, as mentioned in S7, p225, with a marked net advance in the period 1965-1980, coinciding with colder average temperatures over most of the North Atlantic and over western Europe (see S7, graph in Fig 7.12a).

In contrast to the above, high and middle latitude coastal glaciers may grow under warmer and wetter conditions. Mayo and March (1990) report that the Wolverine glacier in the maritime region of southern Alaska near 60°N had generally positive mass balances after 1976 as a result of increased winter precipitation which fell as snow, despite warmer winters, because temperatures remained well below freezing. This change is likely to be associated with the rather striking increase in winter half-year cyclonic activity in the Gulf of Alaska since 1976 discussed in S7, p229. In addition, Fitzharris *et al.* (1992) note that the terminus of the Franz Josef glacier on the west coast of New Zealand has advanced in the 1980s, as have other alpine glaciers in New Zealand, despite the fact that the 1980s were locally one of the warmest decades in the record. Atmospheric circulation changes partly associated with ENSO appear to be involved.

An overall shrinking trend in the glaciers of the Northern Hemisphere has continued into the late 1980s (World Glacier Monitoring Service, 1991). Recent data

(a)

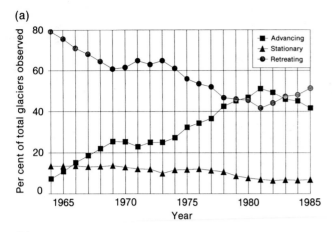

(b)

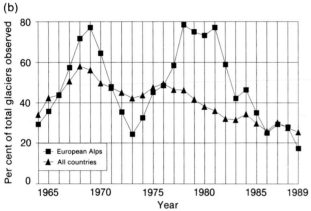

Figure C12: (a) Percentages of observed glacier fronts advancing, stationary and retreating. Data from all available countries; number of glaciers ranges from 271 in 1959/60 to 486 in 1984/5. Values are 5-year moving averages from 1959/60-1963/64 to 1980/81-1984/85. Data from Haeberli and Müller (1988) and Wood (1988); (b) Percentages of observed glaciers having a positive mass balance: (i) all countries (triangles); (ii) about 10 European Alpine glaciers (rectangles). Values are 5-year moving averages from 1959/60-1963/64 (53 glaciers from all countries) to 1984/85-1988/1989 (71 glaciers from all countries). Data from Haeberli and Müller (1988), Wood (1988), World Glacier Monitoring Service (1991), U.S. Geological Survey and Canadian National Hydrology Research Institute.

suggest that mountain glaciers are, as a group, shifting back to a regime dominated by shrinkage and recession, after a period of relative growth and minor re-advance during the 1960s and 1970s (Figures C12a and b). However, this temporary re-advance has not affected the larger glaciers. In further support of this conclusion two detailed regional studies are mentioned. In a study of trends in the length of 242 glaciers in northwest China from the 1950s to 1980s, Shi (1989) shows that 42% were retreating, 29% showed little change and only 29% were advancing. The overall glacier retreat is thought to result from both a warming and a drying trend in this area. In the Southern Hemisphere, Aniya and Naruse (1991) reported that in the northern Patagonian icefield, 20 of the 22 major outlet glaciers were retreating and only one was advancing.

A general review of glacier and other cryospheric trends is given in Barry (1991).

More extensive glacier monitoring and comparative analyses of climatic data are needed to add to these results, as glacier fluctuations are potentially important indicators of regional and global climatic change.

C3.4.3 Polar Ice Caps

Morgan *et al.* (1991) derived time-series of the net rate of snow accumulation since 1806 from four cores separated by up to 700km in a limited area in East Antarctica within 300km of the coast. The four cores yielded fairly similar results. There has been a significant increase in accumulation rate after a minimum around 1960 and the most recent values are the highest observed in the record. Morgan *et al.* (1991) attribute the recent imbalance to increased cyclonic activity around Antarctica leading to higher snowfall. Increased temperatures usually accompany increased snowfall, resulting in positive correlations between temperature and accumulation in Antarctica (Jones *et al.*, 1992). However, it is unclear how widespread this increase in accumulation has been, so it is too soon to revise the estimates of the contribution that Antarctic ice accumulation may make to sea level reduction that were summarized in S9.

The Wordie ice shelf, on the western side of the Antarctic Peninsula, has shrunk markedly from about $2000km^2$ in 1966 to about $700km^2$ in 1989 (Doake and Vaughan, 1991). This is apparently related to a strong regional warming trend. However, the larger ice shelves, such as the Ross and Filchner-Ronne, which help to stabilize the West Antarctic ice sheet, would only be threatened by substantially greater warming (Zwally, 1991).

For evidence from satellites concerning a recent thickening of the Greenland ice sheet, the reader is referred to S9, Section 9.4.4.

C3.4.4 Sea-Ice

No simple relationships have been established between sea-ice extent and temperature, either for the Antarctic (Raper *et al.*, 1984) or for the Northern Hemisphere (Kelly *et al.*, 1987).

A re-analysis of satellite passive microwave data obtained by Nimbus 7 (Gloersen and Campbell, 1991) shows a statistically significant 2.1±0.9% decline in the extent of Arctic sea-ice (including enclosed open water areas) between 1978 and 1987 and an accompanying non-significant 3.5±2.0% reduction in the open-water areas, i.e., polynyas and leads, within the ice field. Figure 7.20a in S7 also indicates a decline of about $0.2 \times 10^6 km^2$ (2%) in the extent of Arctic sea-ice between 1978 and 1987, but there is no decline over the whole period. So the Gloersen and Campbell result, while supported, seems not to be

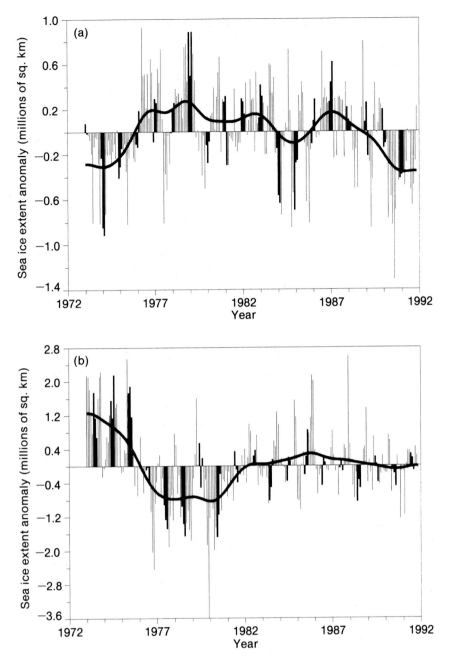

Figure C13: (a) Northern Hemisphere, (b) Southern Hemisphere sea-ice extent anomalies relative to 1973-1991. Data from NOAA (USA). Smooth lines generated from a 39-point binomial filter applied to the monthly anomalies. Heavy bars represent DJF in the NH or JJA in the SH.

representative of the whole period and it cannot therefore be interpreted as a trend. Over the same time interval the above authors found no significant trends in Antarctic sea-ice. Figures C13a and b update the hemispheric sea-ice extent variations shown in Figures 7.20a and b of S7 to the end of 1991. As suggested in S7, no systematic trend toward more or less sea-ice is apparent from the short records compiled to date. In 1990 and 1991, sea-ice extent remained less than the 18-year mean in the Arctic and near the mean in the Antarctic.

It now seems even less likely that the postulated thinning of Arctic sea-ice addressed in S7, based on the submarine observations of Wadhams (1990) north of Greenland, reflects a basin-wide phenomenon. Three issues of concern are (1) the use of data from only 2 cruises in different seasons (May 1987 and October 1976); (2) the year-to-year variability in ice motion in the Beaufort Gyre and Transpolar Drift Stream that may affect the ice regime north of Greenland; (3) the absence of a reliable baseline climatology because the numerous submarine cruises have travelled different routes in different seasons through sectors with widely differing ice conditions (McLaren *et al.*, 1990).

C3.5 Variations and Changes in Atmospheric Circulation

S7, pp225-229, identified in particular recent increases in the intensity of the winter atmospheric circulation over the extratropical Pacific and Atlantic (Figures 7.22 and 7.23): see Trenberth (1990) for a fuller discussion. These findings are reflected in an analysis of the climate of the 1980s by Halpert and Ropelewski (1991) and in analyses by Flohn *et al.* (1992). However, beyond these results, it is difficult to identify new research on atmospheric circulation changes that does not depend critically on how the analyses were done and on the choice of data. It is recognized, though, that studies of atmospheric circulation changes are a key topic for future research as such changes are likely to be an important component of future regional climatic change.

C4 Detection and Attribution of Climatic Change

C4.1 The Greenhouse Effect

"Fingerprint" detection of the influence of an enhanced greenhouse effect, by comparing the patterns of climate change predicted by models with historical observations, is fully discussed in S8. Progress is very difficult (a) because multiple runs of coupled ocean-atmosphere models are needed, requiring very large computer resources; (b) these models may still not be adequate, particularly because of the "flux correction problem" (see Section B2.2); (c) adequate data series are very few and short; and (d) the methodology to implement the technique has not fully matured. In addition, the climate system can respond to many forcings and it remains to be proved that the greenhouse signal is sufficiently distinguishable from other signals to be detected except as a gross increase in tropospheric temperature that is so large that other explanations are not likely (see S8). Because of these difficulties, no definitive paper on detection has appeared since the 1990 IPCC Scientific Assessment. The reader is referred to Karoly *et al.* (1992) for a recent discussion of some of the problems and possible ways forward.

C4.2 Other Factors

C4.2.1 Solar Influences

Possible changes in solar irradiance and their climatic effects were discussed in S2, while Section A2.7 of this Assessment discusses new physically-based evidence about variations in the luminosity of stars like the Sun on century time-scales. Reid (1991) has re-examined the case for solar forcing of global surface temperature in phase with an approximately 80-year ("Gleissberg") modulation of the amplitude of the near 11-year cycle in sunspot numbers. He first formed a relationship between this amplitude and the solar constant using a limited series of measurements since the 1960s. He then used a simple model to predict the global SST from the variation of solar

forcing reconstructed in this way back to 1660, and found that the simulated global SST and a version of the observed global SST were well correlated since 1860. Due to a lack of sufficiently long time-series of satellite measurements of solar irradiance, Reid's conclusions rely considerably on rocket- and balloon-borne measurements of total solar irradiance made in the late 1960s; these are of uncertain accuracy and representativeness. His conclusions also depend on the reliability of the shape of the observed global temperature series, especially prior to 1900 (Reid used an early version of the Bottomley *et al.* (1990) SST data). Strong auto-correlations in both the solar and temperature data also mean that very few degrees of freedom are being compared in each data set.

A new approach by Friis-Christensen and Lassen (1991) (FCL) uses the **length** of the sunspot cycle which is known to be statistically related to solar activity. They hypothesize that shorter cycles are directly related to higher solar magnetic activity which in turn they assume corresponds to a higher total output of solar radiation, though no firm physical basis has been established for this relationship. The length of the solar cycle varies between 7 and 17 years. After smoothing the cycle length data, they find a correlation of -0.95 between cycle length and Northern Hemisphere land surface air temperature since 1860 when averages over corresponding solar cycles of these data sets are compared. FCL do not compare their solar results with a global surface temperature series because they prefer to use temperatures over land which they assume are more likely to be accurately determined than those over the sea. They also find that shorter cycles (supposedly higher solar radiation) correspond with less Icelandic sea-ice since 1740.

In the past, many authors have claimed to largely explain, in a statistical way, the hemispheric or global temperature record in terms of a single forcing factor, but different authors have chosen different factors. A full discussion of such work and why different authors can come to radically different conclusions appears in Wigley *et al.* (1986) and in S2. Recent examples are: (1) After allowing for the influence of ENSO and a (different) index of solar variability, Wu *et al.* (1990) claim that the variations of global and regional night marine air temperature, and to a lesser extent SST, over the last century published by Bottomley *et al.* (1990) can be mainly explained by an atmospheric turbidity index. This index is claimed to relate mainly to the variations in stratospheric dust due to volcanic activity, being based on measurements made at 3100m altitude, but it might, nevertheless, include some influence of tropospheric aerosols as well (Section C4.2.4). (2) Schönwiese and Runge (1991) and Schönwiese and Stahler (1991) explain regional climate changes on the same time-scale in great detail largely in terms of CO_2 forcing, though volcanic, solar and ENSO influences are also included. These and other exercises in curve fitting can demonstrate statistical relationships, but cannot prove physical connections.

A full discussion of the relation between the historical temperature record and different forcing factors requires that greenhouse forcing be included (Kelly and Wigley, 1990; Hansen and Lacis, 1990). Although we do not know what the *response* of the recent past climate to greenhouse forcing has been, we *do* know the approximate *magnitude* ($2Wm^{-2}$ at the top of the troposphere) and *rate of increase* of the greenhouse-gas *forcing* over the last century (Figure 2.2 of S2, p55). Furthermore, empirical studies of relationships between smoothed forcing factors and the statistically non-stationary historical temperature record cannot, alone, resolve the relative contributions of the different forcing factors. The main problem lies in the similarity of the trends of the forcing factors and the small number of degrees of statistical freedom in the data, which often show long-term persistence. In addition, rigorous statistical tools do not exist to show whether relationships between statistically non-stationary data of this kind are truly statistically significant, even though the correlations found by each author between their chosen forcing factor and the temperature record are invariably quite high (Zwiers, personal communication). Thus a **physical** model that includes both the hypothesized forcing and the enhanced greenhouse forcing must be used to make further progress.

C4.2.2 Bi-decadal Temperature Variations

A renewed controversy has developed about the existence of globally significant temperature variations on typically 18-22 year time-scales. These variations are often related to the solar 22-year magnetic cycle (Newell *et al.*, 1989), or sometimes to the 18-19 year lunar orbital cycles (Currie and O'Brien, 1990) which can, in principle, have small tidal effects on the oceans. Ghil and Vautard (1991) have claimed to find a bi-decadal oscillation in global surface air temperature using a novel form of analysis (singular spectrum analysis), mirroring similar findings by Newell *et al.* (1989) in global night marine air temperatures based on more conventional techniques. However, Elsner and Tsonis (1991b) show that the bi-decadal variation which appears in Ghil and Vautard's analysis depends critically on the inclusion of less reliable late nineteenth century data. In view of the possible influence of this variation on our perception of recent trends, further research is warranted.

C4.2.3 "Abrupt" Climatic Changes

Features of many irregular climatic time-series are apparently sudden, usually not very large, changes in average value; or sudden changes of trend, such as are seen in the global and hemispheric temperature series (Figures

C2, C3 and C4). Statistical tests exist to suggest whether the mean has changed significantly (Demarée and Nicolis, 1990) or whether a significant change in trend has occurred (Solow, 1987). Usually, investigators test one or other of these hypotheses, but not both, so the conclusion drawn can depend on the hypothesis that is chosen to be tested and whether tests are chosen in advance of the analysis of the data or subsequently (Karl, 1988). Nevertheless, some of these changes may have a real physical cause and be of obvious importance, such as the rather sudden change in the mean winter half year pressure around 1976 in the extratropical North Pacific shown in Figure 7.23 of S7, p 229 and discussed by Trenberth (1990).

Lin and Zhou (1990) ascribe a climatic "jump" to a series if the difference between the average values for two successive periods is statistically significant according to a t-test, and the length of the transition time between the periods is "much less" than their duration. However, they appear to have pre-selected the dates of the jumps, violating an assumption of the t-test (see Karl, 1988). Also, an analysis is needed of how often such behaviour would occur in truly random series or in stationary series exhibiting persistence. Inevitably, some "jumps" so defined will occur and will be statistical artefacts. The annual Sahel rainfall series illustrates some of the problems (see Figure 7.16b of S7). Demarée and Nicolis (1990) suggest that a statistically significant sudden reduction in average value occurred in 1967-8, whereas the studies of Folland *et al.* (1986) and Rowell *et al.* (1992) indicate that the change of rainfall between the 1950s and 1970s was associated with **a relatively steady, if rapid,** change in interhemispheric SST patterns, a link identified from empirical and modelling studies. It is important for physical understanding of the causes of the Sahel rainfall changes that these contrasting interpretations be reconciled. Many other examples of such "fast" changes exist in regional and local climate records (e.g., Karl and Riebsame, 1984), though few have been subjected to rigorous analysis.

Statistical tests applied to rapid climate changes have a useful role. However, the underlying hypotheses should accord as far as possible with the physical character of the causative processes. Sometimes, for example, apparently abrupt changes between two climatic mean states (e.g., Demarée, 1990) may result from simultaneous processes operating on different time-scales. Climate models should be used to help to uncover these processes.

C4.2.4 Aerosols

C4.2.4.1 Man-made aerosols

A detailed discussion of man-made aerosols, their precursors, and their forcing effects on climate is in Section A2.6. Here we limit discussion to the possible effects of aerosols on the temperature record. The atmospheric concentration of aerosols, particularly those derived from man-made emissions of SO_2, has increased significantly this century in many parts of the Northern Hemisphere, and sulphate concentrations in industrialized areas of eastern Europe and eastern North America are now 10-15 times larger than the expected natural concentration of sulphate (see Section A). The bulk of the increase of SO_2 emissions began in the early 1950s. As stated in Section A, man-made, SO_2-derived aerosols are important because they may have acted to retard the expected warming from the build-up of greenhouse gases and to produce a slower rate of warming in the Northern Hemisphere (i.e., a relative cooling) compared with the Southern Hemisphere. The approximately $1Wm^{-2}$ negative forcing confined to the Northern Hemisphere discussed in Section A2.6.1.2 would offset about 50% of the enhanced greenhouse forcing of about $2Wm^{-2}$ over the past 100 years.

Figures C4a and b show that from about 1910 until the 1950s the Northern Hemisphere surface warmed with respect to that of the Southern Hemisphere. This differential warming, though consistent with some model predictions of more warming in the Northern Hemisphere compared with the Southern Hemisphere as greenhouse gases are added to the atmosphere, is more likely to have been of natural rather than man-made origin, in view of the small increase of these gases over that period. From the 1950s until about 1980, however, the surface of the Southern Hemisphere warmed relative to that of the Northern Hemisphere by nearly 0.3°C (Table C1). The timing of the reversal of relative warming might suggest that sulphate aerosols have been a contributory cause, but renewed warming of the Northern relative to the Southern Hemisphere in the last few years (Figure C4) shows how uncertain such conclusions can be. There is, however, evidence of a reduction in sulphate aerosol emissions from Western Europe since 1980 (Smith, 1991) which could have contributed to renewed warming in the Northern Hemisphere. These results do not prove the hypothesis of a significant aerosol cooling effect on the observed climate because the likely signal is still only of a magnitude similar to the natural variability (Wigley, 1989; Wigley *et al.*, 1992). Many other variables, such as cloudiness changes and ocean circulation variations involving the North Atlantic (S7, Section 7.7), may lead to natural differences between hemispheric temperature trends.

C4.2.4.2 Volcanic aerosols
The major eruption of the Philippines volcano Mt. Pinatubo in June 1991 may be sufficiently great in its climatic effects to allow better physical theories of the influence of volcanic eruptions on climate to be developed. Section A2.6.2 discusses this further. The August 1991

eruption of the Chilean volcano Cerro Hudson was at least 10 times smaller than that of Pinatubo in terms of the amount of SO_2 emitted into the stratosphere, but was still about twice as large as that of Mt. St Helens in 1980 (Doiron *et al.*, 1992). Mt. Pinatubo may have placed nearly 3 times as much material into the stratosphere as did the strong El Chichon eruption of 1982 (Bluth *et al.*, 1992). However, Stowe *et al.* (1992), using a different technique, have estimated that the Pinatubo eruption may have placed a little less than twice as much material there. It is provisionally estimated that this sulphur-rich eruption, which is the type currently thought to have the maximum climatic effects, will temporarily reduce global mean surface temperature by 0.3°-0.5°C over the next year or so (Stowe *et al.*, 1992). The upper end of this estimate is indicated by model simulations carried out by Hansen *et al.* (1992) of the possible effect of Pinatubo stratospheric aerosols on global land temperature, assuming that the aerosols contain twice as much sulphur as did those of El Chichon. If this cooling happens, the consequences of the Pinatubo eruption, with a small extra contribution from Cerro Hudson, could dominate the global surface temperature record in 1992 or a little beyond. In response to heating of the aerosols, globally averaged lower stratospheric temperatures measured by MSU Channel 4 rose by about 1.3°C between June and October 1991, but then began to decline (Figure C9). MSU mid-tropospheric temperatures fell by over 0.5°C between June 1991 and December 1991, though such decreases are common in the record (Spencer and Christy, 1990).

C4.2.5 Changes in Land Surface Characteristics
Balling (1991) has pointed out that changes in the amount of vegetation due to man's activities can change temperatures regionally. Thus, removal of vegetation, perhaps leading eventually to desertification, tends to raise near-surface air temperatures, at least in warmer climates, as more of the incoming solar radiation is used to heat the ground and less is used to provide energy for the transpiration of plants and trees. Such tendencies will be offset by any accompanying increase in surface albedo, and the net effect may depend on the character of the initial vegetation as indicated by studies in Africa (Wendler and Eaton, 1983). Balling contends that, if a fraction of the observed global warming can be attributed to man-made land-use changes, then less can be attributed to the enhanced greenhouse effect. Conversely, such effects need to be balanced against the local surface cooling that is likely in areas where irrigation, new crop growth and new lake surfaces have been developed in recent decades.

C4.2.6. Deep Ocean Heat Storage
A possible cause of the Northern Hemisphere cooling

trend between the 1940s and the 1960s relates to a change in the oceanic circulation in the North Atlantic which temporarily took more heat from the atmosphere into the deep ocean (Wigley and Raper, 1987). S7, p222-223 and Figure 7.18, noted that there is good evidence for warming below about 600m in the North Atlantic between 1957 and 1981 and cooling above this level, especially near the surface. Watts and Morantine (1991) note the evidence for warming in the deeper layers of the North Atlantic and calculate that the heat storage involved would approximately cancel the heat input at the surface due to the simultaneous increase in greenhouse gases, confirming the earlier results of Wigley and Raper (1987).

Recent measurements through the full depth of the ocean in the South-West Pacific at 28°S and 43°S (Bindoff and Church, 1992) show a small warming over much of the water column between 1967 and 1989 which is significantly larger than the seasonal signal or measurement uncertainties. However, the record is too short and intermittent at present to distinguish a warming trend from interdecadal variability.

C5 Concluding Remarks

The main conclusions of S7 concerning the reality, character and magnitude of global and hemispheric warming over the last century remain unaffected by recent evidence. The most challenging result of recent research is probably the increasingly widespread evidence, though not yet conclusive in a globally-averaged sense, of a marked decrease in the diurnal range of temperature over land largely produced by rising minimum temperatures (Section C3.1.5). Is this the result, perhaps partly through increases in cloudiness, of an important anthropogenic effect related to increasing tropospheric aerosols and are increases in greenhouse gases making a contribution? Wider investigations of the response of the diurnal temperature cycle in general circulation models to these effects, differentiated into clear-sky and cloudy conditions, would be very desirable.

In view of the long-standing arguments over the possible role of solar variability in climate change, which have emerged again here, it is desirable that more progress be made in modelling solar dynamics so that likely variations in solar luminosity on time-scales of the solar cycle to millennia can be better judged.

Detecting climate change and attributing it to natural variability, greenhouse forcing or other factors will require a much more **interactive and coordinated** use of data sets and models than has hitherto generally been the case. The historical data base needs much improvement as it has serious problems of heterogeneity (e.g., Parker, 1990), inaccessibility and poor coverage (Folland *et al.*, 1990a; Trenberth *et al.*, 1991). Efforts to tackle some of these problems are now increasing (e.g., Frich *et al.*, 1991). Differences between the satellite MSU data and the radiosonde and surface temperature data need detailed investigation while data describing atmospheric circulation variations, known to contain serious inhomogeneities, needs much more scrutiny. Recent *in situ* and remotely-sensed data need to be optimally blended and made consistent with historical records; this may allow a nearly full global coverage of several data sets for the last decade or more. Heterogeneities in operational analyses of the atmosphere reveal the urgent need for their "re-analysis" over as long a period as possible using a state-of-the-art four-dimensional data assimilation. This will allow the mechanisms of climate change and variability to be much better studied and some of the recent global data sets to be improved.

In the future, remote sensing data must be much more extensively used in climate change studies. However, to achieve the potential of satellite data for studying climate change, the development of methods to ensure homogeneity of processed remote sensing data with historic data will be essential. The satellites, to be truly useful, must provide continual coverage and efforts must be made to sustain their calibration over very long periods. In this context, the proposals for a Global Climate Observing System (GCOS) are highly relevant. In the meantime, surface data from developing countries and from marine archives need to be rescued, quality-controlled and added to existing analyses. Finally, documentation of changing instrumentation, processing schemes, and observing practices must be much better systematized so that artificially induced trends in all climatic data can be more adequately quantified (Karl *et al.*, 1992).

References

Angell, J.K., 1988: Variations and trends in tropospheric and stratospheric global temperatures, 1958-87. *J. Clim.*, **1**, 1296-1313.

Aniya, M. and R. Naruse, 1991: Recent glacier variations in the Patagonia icefields, South America. Abstracts of an International Conference on Climatic Impacts on the Environment and Society (CIES), Tsukuba, Japan, 27 Jan-1 Feb 1991.

Aristarain, A.J., J. Jouzel, and C. Lorius, 1990: A 400-years isotope record of the Antarctic Peninsula Climate. *Geophys. Res. Lett.*, **17**, 2369-2372.

Arkin, P.A., and M. Chelliah, 1990: An assessment of variations of outgoing longwave radiation over the tropics, 1974-1987. Draft manuscript submitted to Section 7 of IPCC 1990.

Balling, R.C., 1991: Impact of desertification on regional and global warming. *Bull. Am. Met. Soc.*, **72**, 232-234.

Barnett, T.P., M.E. Schlesinger and X. Jiang, 1991: On greenhouse gas signal detection strategies. In: *"Greenhouse-Gas-Induced Climatic Change: A Critical Appraisal of Simulations and Observations"*, M.E. Schlesinger (Ed.). Elsevier, Amsterdam, pp537-558.

Barry, R.G., 1991: Observational evidence of changes in global snow and ice cover. In: *"Greenhouse-Gas-Induced Climatic Change: A Critical Appraisal of Simulations and Observations"*, M.E. Schlesinger (Ed.). Elsevier, Amsterdam, pp329-345.

Bates, J.J. and H.F. Diaz, 1991: Evaluation of multi-channel sea surface temperature product quality for climate monitoring: 1982-1988. *J. Geophys. Res.*, **96**, 20613-20622.

Bindoff, N.L. and J.A. Church, 1992: Climate change in the southwest Pacific Ocean? **Nature**. (In press).

Bluth, G.J.S., S.D. Doiron, C.C. Schnetzler, A.J. Krueger and L.S. Walter, 1992: Global Tracking of the SO_2 clouds from the June 1991 Mount Pinatubo eruptions. *Geophys. Res. Lett.*, **19**, 151-154.

Bottomley, M., C.K. Folland, J. Hsiung, R.E. Newell and D.E. Parker, 1990: Global Ocean Surface Temperature Atlas (GOSTA). Joint Met. Office/Massachusetts Institute of Technology Project Project supported by US Dept of Energy, US National Science Foundation and US Office of Naval Research. Publication funded by UK Depts of Energy and Environment. 20+iv pp and 313 Plates. HMSO, London.

Bradley R.S., T. Braziunas, J. Cole, J. Eddy, M. Hughes, J. Jouzel, W. Karlen, K. Kelts, E. Mosley-Thompson, A. Ogilvie, J. Overpeck, J. Pilcher, N. Rutter, M. Stuiver and T.M.L. Wigley, 1991: Global Change: The last 2000 years. Report of Working Group 1. In *"Global Changes of the Past"*. R.S. Bradley, (Ed.). UCAR/Office for Interdisciplinary Earth Studies, Boulder, Colorado, 11-24.

Briffa, K.R., T.S. Bartholin, D. Eckstein, P.D. Jones, W. Karlen, F.H. Schweingruber and P. Zetterberg, 1990: A 1,400-year tree-ring record of summer temperatures in Fennoscandia. *Nature*, **346**, 434-439.

Briffa, K.R., P.D. Jones, T.S. Bartholin, D. Eckstein, F.H. Schweingruber, W. Karlen, P. Zetterberg, and M. Eronen, 1992: Fennoscandian summers from AD 500: temperature changes on short and long timescales. *Clim. Dyn.* (In press).

Brown, B.E., 1990: Coral bleaching. B.E Brown (Ed.). *Coral Reefs*, **8** Special Issue, 153-232.

Bruhl, C. and P.J. Crutzen, 1989: On the disproportionate role of tropospheric ozone as a filter against UV-B radiation. *Geoph. Res. Lett.*, **16**, 296-306.

Bücher, A. and J. Dessens, 1991: Secular trend of surface temperature at an elevated observatory in the Pyrenees. *J. Clim.*, **4**, 859-868.

Cardone, V.J., J.G. Greenwood and M.A. Cane, 1990: On trends in historical marine wind data. *J. Clim.*, **3**, 113-127.

Chelliah, M. and P. Arkin, 1992: Large-scale interannual variability of monthly outgoing long wave radiation anomalies over the global tropics. *J. Clim.*, **5**. (In press).

Chen, L., Y. Shao, M. Dong, Z. Ren and G. Tian, 1991: Preliminary analysis of climatic variation during the last 39 years in China. *Adv. Atmos. Sci.*, **8**, 279-288.

Cook, E., T. Bird, M. Peterson, M. Barbetti, B. Buckley, R. D'Arrigo, R. Francey and P. Tans, 1991: Climate change in

Tasmania inferred from a 1089-year tree-ring chronology of Huon pine. *Science*, **253**, 1266-1268.

Currie, R.G. and D.P. O'Brien, 1990: Deterministic signals in USA precipitation records: Part 1. *Int. J. Climatol.*, **10**, 795-818.

D'Elia, C.F., R.W. Buddemeier and S.V. Smith (Eds.), 1991: Workshop on Coral Bleaching. Coral Reef Ecosystems and Global Change: Report of Proceedings. Maryland Sea Grant College Publication, 51pp.

Demarée, G.R., 1990: Did an abrupt global climatic warming occur in the 1920s? In: *Contributions á l'étude des changements de climat*. Publications IRM, Sér. A, 124, Brussels, 32-37.

Demarée, G.R. and C. Nicolis, 1990: Onset of Sahelian drought viewed as a fluctuation-induced transition. *Q. J. Roy. Met. Soc.*, **116**, 221-238.

Doake, C.S.M. and D.G. Vaughan, 1991: Rapid disintegration of the Wordie Ice Shelf in response to atmospheric forcing. *Nature*, **350**, 328-330.

Doiron S.D., G.J.S. Bluth, C.C. Schnetzler, A.J. Krueger and L.S. Walter, 1992: Transport of Cerro Hudson SO_2 clouds. *EOS Transact.*, **72**, No 45, 489 and 498.

Elliott, W.P. and D.J. Gaffen, 1991: On the utility of radiosonde humidity archives for climate studies. *Bull. Am. Met. Soc.* **72**, 1507-1520.

Elliott, W.P., M.E. Smith and J.K. Angell, 1991: Monitoring tropospheric water vapor changes using radiosonde data. In: *Greenhouse-Gas-Induced Climatic Change: A Critical Appraisal of Simulations and Observations*, M.E. Schlesinger, (Ed.). Elsevier, pp311-328.

Elsner, J.B. and A.A. Tsonis, 1991a: Comparisons of observed Northern Hemisphere surface air temperature records. *Geophys. Res. Lett.*, **18**, 1229-1232.

Elsner, J.B., and A.A. Tsonis, 1991b: Do bi-decadal oscillations exist in the global temperature record? *Nature*, **353**, 551-553.

Farmer, G., T.M.L. Wigley, P.D. Jones and M. Salmon, 1989: Documenting and explaining recent global-mean temperature changes. Climatic Research Unit, Norwich, Final Report to NERC, UK, Contract GR3/6565, 141pp.

Fitzharris, B.B., J.E. Hay and P.D. Jones, 1992: Behaviour of New Zealand glaciers and atmospheric circulation changes over the last 130 years. *Holocene*, **2**, No. 2. (In press).

Flohn, H., A. Kapala, H.R. Knoche and H. Machel, 1990a: Recent changes of the tropical water and energy budget and of mid-latitude circulations. *Clim. Dyn.*, **4**, 237-252.

Flohn, H., A. Kapala, H.R. Knoche and H. Machel, 1990b: Recent changes of the tropical water and energy budget and of mid-latitude circulations. Chaper VI of *Observed Climate Variations and Change: Contributions in support of Section 7 of the 1990 IPCC Scientific Assessment*, D.E. Parker, (Ed.), IPCC/WMO/UNEP, 11pp.

Flohn, H., A. Kapala, H.R. Knoche and H. Machel, 1992: Water vapour as an amplifier of the greenhouse effect: new aspects. *Met. Zeit.*. (In press).

Folland, C.K., 1990: The global and hemispheric surface temperature curves in the IPCC Report. Chapter VII of *Observed Climate Variations and Change: Contributions in Support of Section 7 of the 1990 IPCC Scientific Assessment*, D.E. Parker (Ed.), IPCC/WMO/UNEP, 5pp.

Folland, C.K., 1991: Sea temperature bucket models used to correct historical sea surface temperature data in the Meteorological Office. CRTN 14, 29pp. Available from the National Met. Library, London Rd, Bracknell, Berks, RG12 2SZ, UK.

Folland, C.K., S.J. Foreman, P. Delecluse, K.J. Holmén and P. Gaspar, 1990a: Working Group Report on Climate Observations. In: *Climate-Ocean Interaction.* M.E. Schlesinger (Ed.). Kluwer Academic Publishers, pp353-360.

Folland, C.K., T.R. Karl and K.Ya. Vinnikov, 1990b: Observed climate variations and change. In: *Climate Change, the IPCC Scientific Assessment*, J.T. Houghton, G.J. Jenkins and J.J. Ephraums (Eds.). WMO/UNEP/IPCC, Cambridge University Press, pp195-238.

Folland, C.K., T.N. Palmer and D.E. Parker, 1986: Sahel rainfall and worldwide sea temperatures 1901-85. *Nature*, **320**, 602-607.

Folland, C.K. and D.E. Parker, 1992: A physically-based technique for correcting historical sea surface temperature data. Being submitted to *Q. J. Roy. Met. Soc.*

Folland, C.K., R.W. Reynolds, M. Gordon and D.E. Parker, 1992: A study of six operational sea surface temperature analyses. *J. Clim.* (In press).

Frich, P., B. Brødsgaard and J. Cappelen, 1991: North Atlantic Climatological Dataset (NACD). Present status and future plans. Danish Meteorological Institute - Technical Report 91-8. 27pp.

Friis-Christensen, E. and K. Lassen, 1991: Length of the solar cycle: an indicator of solar activity closely associated with climate. *Science*, **254**, 698-700.

Fu, C. and H.F. Diaz, 1992: An analysis of recent changes in oceanic latent heat flux. Submitted to *J. Clim.*

Gaffen, D.J., T.P. Barnett and W.P. Elliott, 1991: Space and time scales of global tropospheric moisture. *J. Clim.*, **4**, 989-1008.

Gandin, L.S., 1963: Objective analysis of meteorological fields. Translated from Russian by the Israeli Program for Scientific Translations, 1965, 242pp.

Ghil, M. and R. Vautard, 1991: Interdecadal oscillations and the warming trend in global temperature time series. *Nature*, **350**, 324-327.

Gloersen, P. and W.J. Campbell, 1991: Recent variations in Arctic and Antarctic sea-ice covers. *Nature*, **352**, 33-36.

Glynn, P.W., 1991: Coral reef bleaching in the 1980s and possible connections with global warming. *Trends in Ecology and Evolution*, **6**, 175-179.

Glynn, P.W. and W.H. de Weerdt, 1991: Elimination of two reef-building hydrocorals following the 1982-83 El Niño warming event. *Science*, **253**, 69-71.

Groisman, P.Ya., V.V. Koknaeva, T.A. Belokrylova and T.R. Karl, 1991: Overcoming biases of precipitation measurement: a history of the USSR experience. *Bull. Am. Met. Soc.*, **72**, 1725-1733.

Haeberli, W., 1990: Glacier and permafrost signals of 20th-Century warming. *Annals of Glaciology*, **14**, 99-101.

Haeberli, W. and P. Müller, 1988: Fluctuations of glaciers, 1980-1985, Vol. V. World Glacier Monitoring Service, Laboratory of Hydraulics, Swiss Federal Institute of Technology, Zurich. Prepared for the International Association of Hydrological Sciences, UNEP and UNESCO.

Halpert, M.S. and C.F. Ropelewski, 1991: Climate Assessment: A Decadal Review 1981-1990. NOAA/NWS/NMC. 109pp.

Hansen, J.E. and A.A. Lacis, 1990: Sun and dust versus greenhouse gases; an assessment of their relative roles in global climate change. *Nature*, **346**, 713-719.

Hansen, J., A. Lacis, R. Ruedy and M. Sato, 1992: Potential climate impact of Mount Pinatubo eruption. *Geopys. Res. Lett.*, **19**, 215-218.

Hansen, J. and S. Lebedeff, 1988: Global surface air temperatures: update through 1987. *Geophys. Res. Lett.*, **15**, 323-326.

Henderson-Sellers, A., 1990: Review of our current information about cloudiness changes this century. Chapter XI of *Observed Climate Variations and Change: Contributions in support of Section 7 of the 1990 IPCC Scientific Assessment*, D.E. Parker, (Ed.), IPCC/WMO/UNEP, 12pp.

Hulme, M., R. Marsh and P.D. Jones, 1992: Global changes in a humidity index between 1931-60 and 1961-90. *Climate Research.* (Submitted).

IPCC, 1990: *Climate Change: The IPCC Scientific Assessment.* J.T. Houghton, G.J. Jenkins and J.J. Ephraums, (Eds.). Cambridge University Press, Cambridge, UK, 365pp.

Isemer, H-J. and L. Hasse, 1991: The scientific Beaufort equivalent scale: effects on wind statistics and climatological air-sea flux estimates in the North Atlantic Ocean. *J. Clim.*, **4**, 819-836.

Jones, P.D., 1988: Hemispheric surface air temperature variations: recent trends and an update to 1987. *J. Clim.*, **1**, 654-660.

Jones, P.D., 1990: Antarctic temperatures over the present century - a study of the early expedition record. *J. Clim.*, **3**, 1193-1203.

Jones, P.D., P.Ya. Groisman, M. Coughlan, N. Plummer, W.-C. Wang and T.R. Karl, 1990: Assessment of urbanization effects in time series of surface air temperature over land. *Nature*, **347**, 169-172.

Jones, P.D., R. Marsh, T.M.L. Wigley and D.A. Peel, 1992: Decadal timescale links between Antarctic Peninsula ice core oxygen-18 and deuterium and temperature. *Holocene*. (Submitted).

Jones, P.D., T.M.L. Wigley and G. Farmer, 1991: Marine and land temperature data sets: a comparison and a look at recent trends. In: *Greenhouse-Gas-Induced Climatic Change: a Critical Appraisal of Simulations and Observations*. M.E. Schlesinger (Ed). Elsevier, Amsterdam, pp153-172.

Karl, T.R., 1988: Multi-year fluctuations of temperature and precipitation: the gray area of climate change. *Climate Change*, **12**, 179-197.

Karl, T.R. and W.E. Riebsame, 1984: The identification of 10- to 20-year temperature and precipitation fluctuations in the contiguous United States. *J. Clim. and Appl. Met.*, **23**, 950-966.

Karl, T.R., R.R. Heim, Jr. and R.G. Quayle, 1991a: The greenhouse effect in Central North America: If not now when? *Science*, **251**, 1058-1061.

Karl, T.R., G. Kukla, V.N. Razuvayev, M.J. Changery, R.G. Quayle, R.R. Heim, Jr., D.R. Easterling and C.B. Fu, 1991b: Global warming: evidence for asymmetric diurnal temperature change. *Geophys. Res. Lett.*, **18**, 2253-2256.

Karl, T.R., R.G. Quayle and P.Ya Groisman, 1992: Detecting climate variations and change: new challenges for observing and data management systems. *J. Clim.* (In press).

Karoly, D.J., J.A. Cohen, G.A. Meehl, J.F.B. Mitchell, A.H. Oort, R.J. Stouffer and R.T. Wetherald, 1992: An example of fingerprint detection of greenhouse climate change. *Clim. Dyn.* (Submitted).

Kelly, P.M., C.M. Goodess and B.S.G. Cherry, 1987: The interpretation of the Icelandic sea-ice record. *J. Geophys. Res.*, **92**, 10835-10843.

Kelly, P.M., P.D. Jones, T.M.L. Wigley, R.S. Bradley, H.F. Diaz and C.M. Goodess, 1985: The extended Northern Hemisphere surface air temperature record: 1851-1984. In *Extended Summaries, Third Conference on Climate Variations and Symposium on Contemporary Climate: 1850-2100*, American Meteorological Society, pp35-36.

Kelly, P.M. and T.M.L. Wigley, 1990: The influence of solar forcing trends on global mean temperature since 1861. *Nature*, **347**, 460-462.

Labitzke, K. and H. van Loon, 1991: Some complications in determining trends in the stratosphere. *Adv. Space Res.*, **11**, (3)21-(3)30.

Labitzke, K. and H. van Loon, 1992: Association between the 11-year solar cycle and the atmosphere. Part V: *J. Clim.*, **5**. (In press).

Legates, D.R. and C.J. Willmott, 1990: Mean seasonal and spatial variability in gauge-corrected, global precipitation. *Int. J. Climatol.*, **10**, 111-127.

Lin, X.-C. and X. Zhou, 1990: The stages and jumps of long-term variation of the atmospheric circulation. Chapter XIV of *Observed Climate Variations and Change: Contributions in Support of Section 7 of the 1990 IPCC Scientific Assessment*, D.E. Parker (Ed.), IPCC/WMO/UNEP, 17pp.

Lough, J.M., 1991: Rainfall variations in Queensland, Australia, 1891-1986. *Int. J. Climatol.*, **11**, 745-768.

Mayo, L.R. and R.S. March, 1990: Air temperature and precipitation at Wolverine Glacier, Alaska; glacier growth in a warmer, wetter climate. *Ann. Glaciol.*, 191-194.

McClain, E.P., W.G. Pichel and C.C. Walton, 1985: Comparative performance of AVHRR-based multi-channel sea surface temperature. *J. Geophys. Res.*, **90**, 11587-11601.

McLaren, A.S., R.G. Barry and R.H. Bourke, 1990: Could Arctic ice be thinning? *Nature*, **345**, 762.

Meier, M.F., 1984: Contribution of small glaciers to global sea level. *Science*, **226**, 1418-1421.

Morgan, V.I., I.D. Goodwin, D.M. Etheridge and C.W. Wookey, 1991: Evidence from Antarctic ice cores for recent increases in snow accumulation. *Nature*, **354**, 58-60.

Naujokat, B., 1981: Long-term variations in the stratosphere of the Northern Hemisphere during the last two sunspot cycles. *J. Geophys. Res.*, **86**, 9811-9816.

Newell, N.E., R.E. Newell, J. Hsiung and Z. Wu, 1989: Global marine temperature variation and the solar magnetic cycle. *Geophys. Res. Lett.*, **16**, 311-314.

Newell, R.E., 1970: Stratospheric temperature change from the Mt. Agung volcanic eruption of 1963. *J. Atmos. Sci.*, **27**, 977-978.

Nicholls, N. and B. Lavery, 1992: Australian rainfall trends during the twentieth century. *Int. J. Climatol.*, **12**. (In press).

Nitta, T. and S. Yamada, 1989: Recent warming of tropical sea surface temperature and its relationship to the Northern Hemisphere circulation. *J. Met. Soc. Japan*, **67**, 375-383.

Norton, D.A., K.R. Briffa and M.J. Salinger, 1989: Reconstruction of New Zealand summer temperatures to 1730 AD using dendroclimatic techniques. *Int. J. Climatol.*, **9**, 633-644.

Oke, T.R., G.T. Johnson, D.G. Steyn and I.D. Watson, 1991: Simulation of surface urban heat islands under "ideal" conditions at night. Part 2: Diagnostics of causation. *Boundary Layer Met.*, **56**, 339-358.

Oort, A.H. and H. Liu, 1992: Upper air temperature trends over the globe, 1958-1989. *J. Clim.* (In press).

Parker, D.E., 1990: Effects of changing exposure of thermometers at land stations. Chapter XVIII of *Observed Climate Variations and Change: Contributions in Support of Section 7 of the 1990 IPCC Scientific Assessment*, D.E. Parker (Ed.), IPCC/WMO/UNEP, 31pp.

Parker, D.E. and P.D. Jones, 1991: Global warmth in 1990. *Weather*, **46**, 302-311.

Parker, D.E., T. Legg and C.K. Folland, 1992: A new daily Central England Temperature series, 1772-1991. *Int. J. Climatol.* (In press).

Patzelt, G. and M. Aellen, 1990: Gletscher. In: *Schnee, Eis und Wasser der Alpen in einer wärmeren Atmosphäre*. D. Vischer, (Ed.). *Versuchsansalt für Wasserbau, Hydrologie und Glaziologie*, **108**, 49-69. ETH, Zurich.

Peel, D.A., 1992: Ice core evidence from the Antarctic Peninsula region. In: *Climate Since AD1500*, R.S. Bradley and P.D. Jones (Eds.). Routledge, pp.549-571.

Plantico, M.S., T.R. Karl, G. Kukla and J. Gavin, 1990: Is recent climate change across the United States related to rising levels of anthropogenic greenhouse gases? *J. Geophys. Res.*, **95**, 16,617-16,637.

Plummer, N., 1991: Annual mean temperature anomalies over eastern Australia. *Bull. Austr. Met. and Ocean. Soc.*, **4**, 42-44.

Quayle, R.G., D.R. Easterling, T.R. Karl and P.M. Hughes, 1991: Effects of recent thermometer changes in the cooperative station network. *Bull. Am. Met. Soc.*, **72**, 1718-1723.

Raper, S.C.B., T.M.L. Wigley, P.R. Mayes, P.D. Jones and M.J. Salinger, 1984: Variations in surface air temperatures. Part 3: The Antarctic, 1957-82. *Monthly Weather Review*, **112**, 1341-1353.

Reid, G.C., 1991: Solar total irradiance variations and the global sea surface temperature record. *J. Geophys. Res.*, **96**, 2835-2844.

Reynaud, L., M. Vallon, S. Martin and A. Letreguilly, 1984: Spatio-temporal distribution of the glacial mass balance in the Alpine, Scandinavian and Tien Shan areas. *Geografiska Annaler*, **66A/3**, 239-247.

Reynolds, R.W., C.K. Folland and D.E. Parker, 1989: Biases in satellite-derived sea surface temperature data. *Nature*, **341**, 728-731.

Reynolds R.W. and D.C. Marsico, 1992: An improved real-time global sea surface temperature analysis. *J. Clim.* (Submitted).

Roberts, L., 1991: Greenhouse role in reef stress unproven. *Science*, **253**, 258-259.

Robinson, D.A., F.T. Keimig and K.F. Dewey, 1991: Recent variations in Northern Hemisphere snow cover. Proc. 15th NOAA Annual Climate Diagnostics Workshop, Asheville, N.C., 29 Oct-2 Nov 1990, pp219-224.

Rowell D.P., C.K. Folland, K. Maskell, J.A. Owen and M.N. Ward, 1992: Causes and predictability of Sahel rainfall variability. *Geophys. Res. Lett.* (In press).

Salinger, M.J., 1990: Southern climates in the 19th and 20th centuries. In: *Southern Landscapes*, G. Kearsley and B. Fitzharris (Eds.). Univ. of Otago, Dunedin, pp27-38.

Salinger, M.J. and B. Collen, 1991: Climate trends in the South Pacific. Proceedings of the Conference *South Pacific Environments: Interactions with weather and climate*, University of Auckland, Auckland, New Zealand, pp176-177.

Schönwiese, C.-D. and K. Runge, 1991: Some updated statistical assessments of the surface temperature response to increased greenhouse gases. *Int. J. Climatol.*, **11**, 237-250.

Schönwiese, C.-D. and U. Stahler, 1991: Multi-forced statistical assessments of greenhouse-gas-induced surface air temperature change 1890-1985. *Clim. Dynam.*, **6**, 23-33.

Scotto, J., G. Cotton, F. Urbach, D. Berger and T. Fears, 1988: Biologically-effective ultraviolet radiation: Surface measurements in the United States, 1974-1985. *Science*, **239**, 762-764.

Shi, Y.F., 1989: The warming and dry trend in Northwest China according to the changes of mountain glaciers and the lakes. Proc. Water Resources Conf., Dept of Earth Sciences, Chinese Academy of Sciences, Beijing, Jan 1989.

Shine, K.P., R.G. Derwent, D.J. Wuebbles and J-J. Morcrette, 1990: Radiative forcing of climate. In: *Climate Change. The IPCC Scientific Assessment*, J.T. Houghton, G.J. Jenkins and J.J. Ephraums (Eds.). WMO/UNEP/IPCC, Cambridge University Press, pp41-68.

Shine, K.P. and A. Sinha, 1991: Sensitivity of the Earth's climate to height-dependent changes in the water vapour mixing ratio. *Nature*, **354**, 382-384.

Smith, F.B., 1991: Regional air pollution, with special emphasis on Europe. *Q.J. Roy. Met. Soc.*, **117**, 657-683.

Solow, A.R., 1987: Testing for climate change: an application of the two phase regression model. *J. Clim. Appl. Met.*, 26, 1401-1405.

SORG, 1991: *Stratospheric Ozone 1991*. Report of the UK Stratospheric Ozone Review Group. HMSO, London, 18pp.

Spencer, R.W. and J.R. Christy, 1990: Precise monitoring of global temperature trends from satellites. *Science*, **247**, 1558-1562.

Spencer, R.W. and J.R. Christy, 1992a: Precision and radiosonde validation of satellite gridpoint temperature anomalies, Part I: MSU Channel 2. *J. Clim.* (In press).

Spencer, R.W. and J.R. Christy, 1992b: Precision and radiosonde validation of satellite gridpoint temperature anomalies, Part II: A tropospheric retrieval and trends 1979-90. *J. Clim.* (In press).

Spencer, R.W., J.R. Christy and N.C. Grody, 1990: Global atmospheric temperature monitoring with satellite microwave measurements: method and results 1979-84. *J. Clim.*, **3**, 1111-1128.

Stowe, L.L., R.M. Carey and P.P. Pellegrino, 1992: Monitoring the Mt. Pinatubo aerosol layer with NOAA/11 AVHRR data. *Geophys. Res. Lett.*, **19**, 159-162.

Strong, A.E., 1989: Greater global warming revealed by satellite-derived sea surface temperature trends. *Nature*, **338**, 642-645.

Trenberth, K.E., 1990: Recent observed interdecadal climate changes in the Northern Hemisphere. *Bull. Amer. Met. Soc.*, **71**, 988-993.

Trenberth, K.E., J.R. Christy and J.W. Hurrell, 1992: Monitoring global monthly mean surface temperatures. *J. Clim.* (In press).

Trenberth, K.E. and J.G. Olson, 1989: Temperature trends at the South Pole and McMurdo Sound. *J. Clim.*, **2**, 1196-1206.

Trenberth, K.E. and J.G. Olson, 1991: Representativeness of a 63-station network for depicting climate changes. In *Greenhouse-Gas-Induced Climatic Change: A Critical Appraisal of Simulations and Observations*, M.E. Schlesinger, (Ed.). Elsevier, pp249-259.

Trenberth, K.E., & 16 other authors, 1991: Working Group 1: Climate Observations. In: *Greenhouse-Gas-Induced Climatic Change: A Critical Appraisal of Simulations and Observations*, M.E. Schlesinger (Ed.), Elsevier, Amsterdam, pp571-582.

Van Engelen, A.F.V. and J.W. Nellestijn, 1992: Monthly, seasonal and annual means of air temperature in tenths of centigrade in De Bilt, Netherlands, 1706-1991. KNMI, De Bilt, 5pp.

Van Loon, H. and K. Labitzke, 1990: Association between the 11-year solar cycle and the atmosphere. Part IV: The stratosphere, not grouped by the phase of the QBO. *J. Clim.*, **3**, 827-837.

Vinnikov, K.Ya., P.Ya. Groisman and K.M. Lugina, 1990: Empirical data on contemporary global climate changes (temperature and precipitation). *J. Clim.*, **3**, 662-677.

Wadhams, P., 1990: Evidence for thinning of the Arctic sea ice cover north of Greenland. *Nature*, **345**, 795-797.

Wang, S.-W. and R. Wang, 1991: Little Ice Age in China. *Chinese Sci. Bull.*, **36**, 217-220.

Wang, R., S.-W. Wang and K. Fraedrich, 1991: An approach to reconstruction of temperature on a seasonal basis using historical documents from China. *Int. J. Climatol.*, **11**, 381-392.

Ward, M.N., 1992: Provisionally corrected surface wind data, worldwide ocean-atmosphere surface fields and Sahel rainfall variability. *J. Clim.* (In press).

Warrick, R. and J. Oerlemans, 1990: Sea level rise. In: *Climate Change. The IPCC Scientific Assessment*, J.T. Houghton, G.J. Jenkins and J.J. Ephraums (Eds.). WMO/UNEP, IPCC, Cambridge University Press, Cambridge, UK, pp257-281.

Watts, R.G. and M.C. Morantine, 1991: Is the greenhouse gas climate signal hiding in the deep ocean? *Climatic Change*, **18**, No. 4, ppiii-vi.

Weber, G.-R., 1990: Spatial and temporal variation of sunshine in the Federal Republic of Germany. *Theor. Appl. Clim.*, **41**, 1-9.

Wendler, G. and F. Eaton, 1983: On the desertification of the Sahel zone. Part I. Ground observations. *Climatic Change*, **5**, 365-380.

Wigley, T.M.L., 1989: Possible climate change due to SO_2 derived cloud condensation nuclei. *Nature*, **339**, 365-367.

Wigley, T.M.L. and T.P. Barnett, 1990: Detection of the greenhouse effect in the observations. In *Climate Change: The IPCC Scientific Assessment*, J.T. Houghton, G.J. Jenkins and J.J. Ephraums (Eds.). WMO/UNEP/IPCC, Cambridge University Press, pp239-255.

Wigley, T.M.L., P.D. Jones and P.M. Kelly, 1986: Empirical Climate Studies. In: *The Greenhouse Effect, Climatic Change and Ecosystems*, B. Bolin, B.R. Döös, J. Jäger and R.A. Warrick (Eds.). SCOPE 29, Wiley, Chichester.

Wigley, T.M.L., P.D. Jones, P.M. Kelly and M. Hulme, 1992: Update of global surface temperature changes and an interpretation of past variations. Proc. 16th NOAA Annual Climate Diagnostics Workshop. (In press).

Wigley, T.M.L. and S.C.B. Raper, 1987: Thermal expansion of seawater associated with global warming. *Nature*, **330**, 127-131.

Williams, R.S., Jr. and J.G. Ferrigno, (Eds.), 1988-1991: Satellite image atlas of glaciers of the world. U.S. Geological Survey Professional Papers 1386-B "Antarctica" (1988); 1386-G "Middle East and Africa" (1991); 1386-H "Irian Jaya, Indonesia and New Zealand" (1989). Washington, D.C.

WMO, 1991: Global Energy and Water Cycle Experiment (GEWEX). Report of the Third Session of the JSC Scientific Steering Group for GEWEX. ICSU/WMO WCRP-57, WMO/TD No. 424, 63pp.

Wood, F.B., 1988: Global Alpine glacier trends, 1960's - 1980's. *Arctic and Alpine Research*, **20**, 404-413.

Woodruff, S.D., 1990: Preliminary comparison of COADS(US) and MDB(UK) ship reports. Chapter XXVII of *Observed Climate Variations and Change: Contributions in Support of Section 7 of the 1990 IPCC Scientific Assessment*, D.E. Parker (Ed.), IPCC/WMO/UNEP, 36pp.

Woodruff, S.D., R.J. Slutz, R.L. Jenne and P.M. Steurer, 1987: A Comprehensive Ocean-Atmosphere Data Set. *Bull. Amer. Met. Soc.*, **68**, 1239-1250.

World Glacier Monitoring Service, 1991: Glacier Mass Balance Bulletin No. 1 (1988-1989). A contribution to the Global Environment Monitoring System (GEMS) and the International Hydrological Programme, W. Haeberli and E. Herren (Eds.). IAHS (ICSU)/UNEP/ UNESCO, 70pp.

World Water Balance and Water Resources of the Earth, 1974 (Russian) and 1978 (English). UNESCO, 638pp.

Wu, Z., R.E. Newell and J. Hsiung, 1990: Possible factors controlling global marine temperature variations over the past century. *J. Geophys. Res.*, **95**, 11799-11810.

Zwally, H.J., 1991: Global warming: Breakup of Antarctic ice. *Nature*, **350**, 274.

ANNEX

Climatic consequences of emissions and a comparison of IS92a and SA90

J.F.B. Mitchell; J.M. Gregory

Annex

For the 1990 IPCC Scientific Assessment (IPCC, 1990), IPCC Working Group III developed four scenarios of future emissions of greenhouse gases. These were used by Working Group I to produce estimates of the resulting rate and magnitude of climate change, represented by the global average surface temperature. In the current report, six revised emissions scenarios (referred to as IS92a-f) have been described (see Section A3). The purpose of this Annex is to present calculations of the temperature rise consequent on these new scenarios and to compare IS92a with the IPCC 1990 Scenario A (SA90) using the same assumptions as those used in IPCC (1990).

The calculations are performed using a one-dimensional climate model and a series of gas cycle models that relate emissions to concentration changes (STUGE, Wigley *et al.*, 1991). The climate model is essentially the same as the model used by IPCC (1990) and it reproduces their results well. For instance, using SA90 with a climate sensitivity of 2.5°C, IPCC 1990 found a warming of 3.3°C over the period 1990-2100 (see IPCC, 1990, Figure A.9). For the same scenario, the model gives the same figure to within 0.1°C. The gas cycle models simulate well the IPCC (1990) concentration projections (which were obtained by averaging the results from a number of different models) for the full range of emission scenarios considered by IPCC (1990).

The inputs to the model are the projected emissions of carbon dioxide (CO_2), methane (CH_4), nitrous oxide (N_2O) and the halocarbons CFC-11, CFC-12 and HCFC-22. CO_2 emissions are converted to concentrations using a simple carbon cycle model, developed by parametrizing results from an ocean circulation model including a carbon cycle. CH_4 concentrations are obtained from a mass balance with exponential decay terms representing atmospheric destruction and the soil sink. The lifetime for the former process is made to depend on CH_4 concentration and CO_2 emissions. N_2O and halocarbons, however, are treated as having constant decay times. Note that we have not updated the atmospheric lifetimes of CFC-11 (65 years), CFC-12 (130 years) and CH_4 to the later estimates given in Section A3.

The time-series of atmospheric concentration of the various gases are used to calculate the resulting changes in radiative forcing, following Table 2.2 of IPCC (1990). Halocarbons, other than the three named above, are not handled directly by the model, but are accounted for by scaling up, by a factor of 1.43, the combined effects of CFC-11 and CFC-12. The direct radiative effect of the CH_4 concentration is also inflated, by a factor of 1.3, in order to represent the effect of stratospheric water vapour produced by oxidation of methane. No account is taken of the influence of ozone or sulphate aerosols.

The time-series of radiative forcing is then applied to an upwelling-diffusion energy-balance climate model. Essentially, this comprises an oceanic mixed-layer coupled to a vertical water column in which heat is transported by diffusion and advection. The model is one-dimensional, but the presence of land masses is accounted for by having separate, zero-heat-capacity boxes for land in each hemisphere. The response of the climate system to the changes in radiative forcing is principally determined by the climate sensitivity, normally quoted as the equilibrium global mean temperature change expected for a doubling of atmospheric CO_2. In the model, this parameter accounts for all ways in which the mixed-layer loses heat except for the flux into the deep ocean. Climate feedbacks are not explicitly modelled but their effects are accounted for by the choice of a range of values of the climate sensitivity. Three different values were used, namely 1.5°C, 2.5°C (the IPCC (1990) "best-estimate") and 4.5°C. For these and the other parameters of the model (depth of the mixed-layer, ocean thermal diffusivity, rate of upwelling and temperature response of polar sinking water) the same values were chosen as in IPCC (1990).

Two points should be noted concerning the emissions scenarios as they are entered into the model. Firstly, the model fixes the 1990 emissions in each case to accord with the values used in IPCC (1990) . These figures do not agree with those in IS92; this causes an inevitable discrepancy between the intended emissions and the figures actually used for 1990-1995 (for the halocarbons) or 1990-2000 (for the other gases). Secondly, the current version of the model requires scenarios to be specified as figures at a certain number of fixed times, and linear interpolation is performed between them. Thus the points used by the model do not all correspond to those used in the scenarios, leading to small differences in some cases because of discrepancies in the periods of linear interpolation.

The largest radiative forcing is produced by IS92e and f (Figure Ax.1). CFCs are phased out more rapidly in IS92e, giving slightly less forcing early in the century, but larger increases in CO_2 emissions later on result in larger forcing in IS92e. Of the middle scenarios (IS92a and b), IS92a has slightly greater CO_2 emissions, while CFCs are phased out faster in IS92b, so the forcing is larger in IS92a throughout. In the low scenarios (IS92c and d), CFCs are phased out rapidly in IS92d early on, giving smaller forcing, but large reductions in CO_2 emissions in IS92c later on reverse the order of the magnitudes of forcing in the latter half of the century.

Using the "best-estimate" sensitivity, these scenarios give a range of warming from 1.5 to 3.5°C by the year 2100 (Figure Ax.2). Also shown are the warmings with low, best-estimate and high climate sensitivities for IS92a which give warmings of 2 to 4°C over the next hundred years (Figure Ax.3).

It is also instructive to compare IS92a with SA90. The concentrations of CO_2 and N_2O are similar (Figures Ax.4 and Ax.5) but CH_4 concentrations are slightly higher in the 1990 Scenario. The difference in forcing due to these gases is about 0.4Wm^{-2} (Figure Ax.6). However, the concentrations of HCFC grow much more rapidly in the SA90 case, and the concentrations of CFCs increase whereas in IS92a they decrease (Figure Ax.7) giving a difference of about 0.7Wm^{-2} in radiative forcing (Figure Ax.8). Thus the forcing is about 15% less in IS92a (Figure Ax.9), giving a slightly smaller warming (Figure Ax.10).

References:

IPCC, 1990: *Climate Change: The IPCC Scientific Assessment.* J.T. Houghton, G.J. Jenkins and J.J. Ephraums (Eds.). Cambridge University Press, Cambridge, UK, 365pp.

Wigley, T.M.L., T. Holt and S.C.B. Raper, 1991: STUGE (An Interactive Greenhouse Model): User's Manual. Climatic Research Unit, Norwich, UK, 44pp.

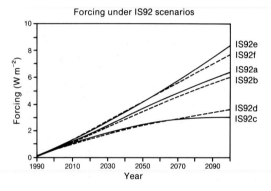

Figure Ax.1: Changes (post-1990) in radiative forcing (Wm^{-2}) arising from scenarios IS92a-f derived from the model of Wigley *et al.* (1991). IS92a, c and e are shown as solid curves; b, d and f are shown as dashed. No allowance has been made for changes in ozone or for the effects of increases in sulphate aerosol.

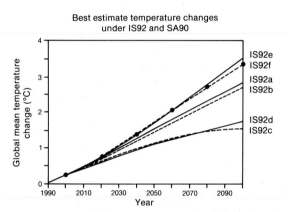

Figure Ax.3: As for Figure Ax.1 but for temperature change (°C) assuming a "Best-estimate" climate sensitivity of 2.5°C. The changes due to SA90 are shown as solid circles.

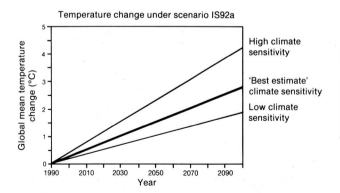

Figure Ax.2: Simulated changes in global mean temperature after 1990 under IS92a due to doubling CO_2, assuming High, "Best-estimate" and Low climate sensitivities (4.5, 2.5 and 1.5°C respectively).

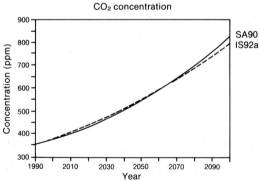

Figure Ax.4: CO_2 concentrations (ppm) derived from IS92a (dashed curve) and SA90 (solid curve).

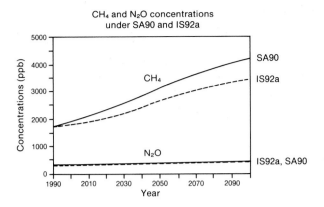

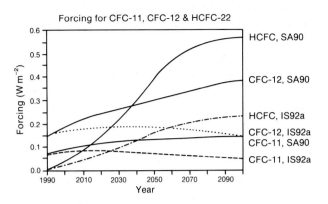

Figure Ax.5: As for Figure Ax.4 but for methane and nitrous oxide. IS92a = dashed curves; SA90 = solid curves.

Figure Ax.8: Changes (since pre-industrial times) in radiative forcing (Wm^{-2}) due to the changes in halocarbon concentrations shown in Figure Ax.7. IS92a = dotted and dashed curves; SA90 = solid curves.

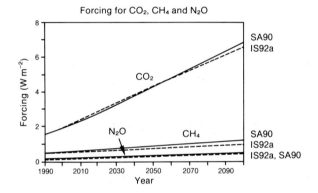

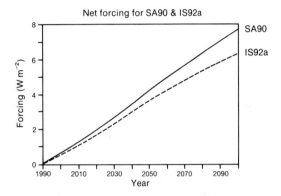

Figure Ax.6: Changes (since pre-industrial times) in radiative forcing (Wm^{-2}) due to increases in carbon dioxide, methane and nitrous oxide. IS92a = dashed curves; SA90 = solid curves.

Figure Ax.9: Changes (post-1990) in net radiative forcing under SA90 and IS92a (all gases). No allowance has been made for changes in ozone concentrations or for increases in sulphate aerosols. IS92a = dashed curve; SA90 = solid curve.

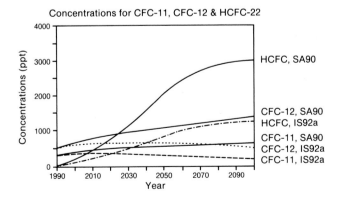

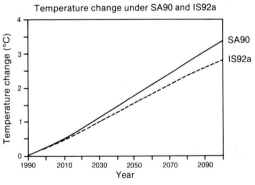

Figure Ax.7: Changes in concentrations (ppt) of HCFC-22, CFC-11 and CFC-12. IS92a = dotted and dashed lines; SA90 = solid curves.

Figure Ax.10: As Figure Ax.9 but showing changes of mean global temperature after 1990 assuming a "Best-estimate" climate sensitivity of 2.5°C. IS92a = dashed curve; SA90 = solid curve.

Appendix 1

ORGANIZATION OF IPCC AND WORKING GROUP I

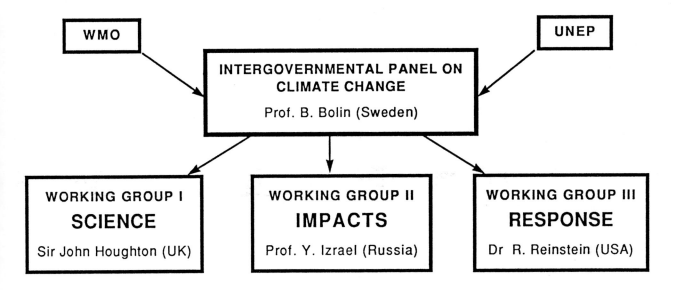

IPCC:

Chairman:	Professor B. Bolin *(Sweden)*
Vice Chairman:	Dr A. Al Gain *(Saudi Arabia)*
Rapporteur:	Dr J.A. Adejokun *(Nigeria)*
Secretary:	Dr N. Sundararaman *(WMO)*

Working Group I:

Chairman:	Sir John Houghton *(United Kingdom)*
Vice Chairmen:	Dr M. Seck *(Senegal)*
	Dr A.D. Moura *(Brazil)*
	Dr M. Sanwal *(India)*
	Dr H. Grassl *(Germany)*

Core Team at the UK Meteorological Office:

Coordinator:	Dr B.A. Callander
Technical Editor:	Miss S.K. Varney
Sub-Editor:	Mr J.J. Ephraums
Visiting Scientist:	Dr R.T. Watson *(USA)*

Appendix 2

CONTRIBUTORS TO THE IPCC WGI REPORT

SECTION A

Overall Coordinating Author:
R.T. Watson   NASA Headquarters, USA

SECTION A1

Lead Authors:
R.T. Watson   NASA Headquarters, USA
L.G. Meira Filho   Instituto Nacional de Pesquisas Espaciais, Brazil
E. Sanhueza   Instituto Venezolano de Investigaciones Cientificas, Venezuela
A. Janetos   NASA Headquarters, USA

Contributors:
P.J. Fraser [†]   CSIRO Division of Atmospheric Research, Australia
I.J. Fung   NASA Goddard Space Flight Center, USA
M.A.K. Khalil   Oregon Graduate Institute of Science and Technology, USA
M. Manning   DSIR Institute of Nuclear Sciences, New Zealand (currently at NOAA, USA)
A. McCulloch   ICI Chemicals and Polymers, UK
A.P Mitra   National Physical Laboratory, India
B. Moore   University of New Hampshire, USA
H. Rodhe   Stockholm University, Sweden
D. Schimel   Colorado State University, USA
U. Siegenthaler   Physics Institute, University of Bern, Switzerland
D. Skole   University of New Hampshire, USA
R.S. Stolarski [†]   NASA Goddard Space Flight Center, USA

[†] - *Chairs for Chapters 1 and 2 of WMO/UNEP Science Assessment of Ozone Depletion: 1991 (WMO, 1992), therefore representing a number of others who contributed to this Section but who are not listed here.*

SECTION A2

Lead Authors:
I.S.A. Isaksen   Institute of Geophysics, University of Oslo, Norway
V. Ramaswamy   NOAA Geophysical Fluid Dynamics Laboratory, USA
H. Rodhe   Stockholm University, Sweden
T.M.L. Wigley   Climatic Research Unit, University of East Anglia, UK

Contributors:
B.A. Callander   Meteorological Office, UK

R. Charlson	University of Washington, USA
I. Karol	Main Geophysical Observatory, Russia
J. Lelieveld	Max-Planck-Institute for Chemistry, Germany
C.B. Leovy	University of Washington, USA
S.E. Schwartz	Brookhaven National Laboratories, USA
K.P. Shine	University of Reading, UK
R.T. Watson	NASA Headquarters, USA
D.J. Wuebbles	Lawrence Livermore National Laboratory, USA

SECTION A3

Lead Authors:

J. Leggett	Environmental Protection Agency, USA
W.J. Pepper	ICF Information and Technology, Inc., USA
R.J. Swart	RIVM, The Netherlands

Contributors:

J.A. Edmonds	Battelle Pacific Northwest Laboratories (DOE), USA
L.G. Meira Filho	Instituto Nacional de Pesquisas Espaciais, Brazil
I. Mintzer	University of Maryland, USA
M.-X. Wang	Institute of Atmospheric Physics, Academy of Sciences, China
J. Wasson	ICF Incorporated, USA

SECTION B

Lead Authors:

W.L. Gates	Lawrence Livermore National Laboratory, USA
J.F.B. Mitchell	Hadley Centre for Climate Prediction & Research, Meteorological Office, UK
G.J. Boer	Canadian Climate Center, Canada
U. Cubasch	Max-Planck-Institut für Meteorologie, Germany
V.P. Meleshko	Main Geophysical Observatory, Russia

Contributors:

D. Anderson	Hooke Institute for Atmospheric Research, UK
W. Broecker	Lamont-Doherty Geological Observatory, USA
D. Cariolle	Météo-France, Centre National de Recherches Météorologiques, France
H. Cattle	Hadley Centre for Climate Prediction & Research, Meteorological Office, UK
R.D. Cess	State University of New York, USA
F. Giorgi	National Center for Atmospheric Research, USA
M.I. Hoffert	New York University, USA
B.G. Hunt	CSIRO Division of Atmospheric Research, Australia
A. Kitoh	Meteorological Research Institute, Japan
P. Lemke	Alfred-Wegener Institute for Polar & Marine Research, Germany
H. Le Treut	Laboratoire de Météorologie Dynamique du CNRS, France
R.S. Lindzen	Massachusetts Institute of Technology, USA
S. Manabe	NOAA Geophysical Fluid Dynamics Laboratory, USA
B.J. McAvaney	Bureau of Meteorology Research Centre, Australia
L. Mearns	National Center for Atmospheric Research, USA
G.A. Meehl	National Center for Atmospheric Research, USA
J.M. Murphy	Hadley Centre for Climate Prediction & Research, Meteorological Office, UK
T.N. Palmer	European Centre for Medium Range Weather Forecasts, UK
A.B. Pittock	CSIRO Division of Atmospheric Research, Australia
K. Puri	Bureau of Meteorology Research Centre, Australia
D.A. Randall	Colorado State University, USA
D. Rind	NASA Goddard Institute for Space Studies, USA
P.R. Rowntree	Hadley Centre for Climate Prediction & Research, Meteorological Office, UK

M.E. Schlesinger	University of Illinois, USA
C.A. Senior	Hadley Centre for Climate Prediction & Research, Meteorological Office, UK
I.H. Simmonds	University of Melbourne, Australia
R. Stouffer	NOAA Geophysical Fluid Dynamics Laboratory, USA
S. Tibaldi	Università di Bologna, Italy
T. Tokioka	Meteorological Research Institute, Japan
G. Visconti	Università degli Studi Dell'Aquila, Italy
J.E. Walsh	University of Illinois, USA
W.-C. Wang	State University of New York, USA
D. Webb	Institute of Oceanographic Sciences, UK

SECTION C

Lead Authors:

C.K. Folland	Hadley Centre for Climate Prediction & Research, Meteorological Office, UK
T.R. Karl	NOAA National Climate Data Center, USA
N. Nicholls	Bureau of Meteorology, Australia
B.S. Nyenzi	Directorate of Meteorology, Tanzania
D.E. Parker	Hadley Centre for Climate Prediction & Research, Meteorological Office, UK
K.Ya. Vinnikov	State Hydrological Institute, Russia (currently at GFDL, USA)

Contributors:

J.K. Angell	NOAA Air Resources Laboratory, USA
R.C. Balling	Arizona State University, USA
R.G. Barry	University of Colorado, USA
M. Chelliah	NOAA Climate Analysis Center, USA
L. Chen	State Meteorological Administration, China
J.R. Christy	University of Alabama, USA
G.R Demarée	Royal Meteorological Institute, Belgium
H.F. Diaz	NOAA Environmental Research Laboratories, USA
Y. Ding	State Meteorological Administration, China
W.P. Elliott	NOAA Air Resources Laboratory, USA
H. Flohn	Meteorological Institute, Germany
E. Friis-Christensen	Danish Meteorological Institute, Denmark
C. Fu	Institute of Atmospheric Physics, Academy of Sciences, China
P.Ya Groisman	State Hydrological Institute, Russia
W. Haeberli	Versuchsanstalt für Wasserbau, Switzerland
J.E. Hansen	NASA Goddard Institute for Space Studies, USA
P.D. Jones	Climatic Research Unit, University of East Anglia, UK
D.J. Karoly	Monash University, Australia
K. Labitzke	Free University of Berlin, Germany
K. Lassen	Danish Meteorological Institute, Denmark
P.J. Michaels	University of Virginia, USA
A.H. Oort	NOAA Geophysical Fluid Dynamics Laboratory, USA
R.W. Reynolds	NOAA Climate Analysis Center, USA
A. Robock	University of Maryland, USA
C.F. Ropelewski	NOAA Climate Analysis Center, USA
M.J. Salinger	Meteorological Service, New Zealand
R.W. Spencer	NASA Marshall Space Flight Center, USA
A.E. Strong	Center for Excellence in Oceanic Research, The Navy Academy, USA
K.E. Trenberth	National Center for Atmospheric Research, USA
S. Tudhope	University of Edinburgh, UK
S.-W. Wang	Beijing University, China
M.N. Ward	Hadley Centre for Climate Prediction & Research, Meteorological Office, UK
T.M.L. Wigley	Climatic Research Unit, University of East Anglia, UK

| H. Wilson | NASA Goddard Institute for Space Studies, USA |
| F.B. Wood | US Congress, Office of Technology Assessment, USA |

ANNEX

Lead Authors:

| J.F.B. Mitchell | Hadley Centre for Climate Prediction & Research, Meteorological Office, UK |
| J.M. Gregory | Hadley Centre for Climate Prediction & Research, Meteorological Office, UK |

Appendix 3

REVIEWERS OF THE IPCC WGI REPORT

The persons named below all contributed to the peer review of the IPCC Working Group I 1992 Supplement and its supporting scientific evidence. Whilst every attempt was made by the Lead Authors to incorporate their comments, in some cases these formed a minority opinion which could not be reconciled with the larger consensus. Therefore, there may be persons below who still have points of disagreement with areas of the Report.

ARGENTINA

M.C. Massei — Area de Evaluacion Ambiental y Social

AUSTRALIA

R. Allan	CSIRO (Division of Atmospheric Research)
W. Bouma	CSIRO (Division of Atmospheric Research)
W.F. Budd	University of Tasmania
P. Cheng	Department of the Arts, Sport, the Environment & Territories
J. Church	CSIRO (Division of Oceanography)
R. Colman	Bureau of Meteorology Research Centre
M.J. Coughlan	Bureau of Meteorology (currently at NOAA Office of Global Programs, USA)
M. Dix	CSIRO (Division of Atmospheric Research)
I. Enting	CSIRO (Division of Atmospheric Research)
J. Evans	CSIRO (Division of Atmospheric Research)
P.M. Fleming	CSIRO (Division of Water Resources)
R.J. Francey	CSIRO (Division of Atmospheric Research)
P.J. Fraser	CSIRO (Division of Atmospheric Research)
J. Fraser	Bureau of Meteorology Research Centre
J.S. Frederiksen	CSIRO (Division of Atmospheric Research)
I.E. Galbally	CSIRO (Division of Atmospheric Research)
D.J. Gauntlett	Bureau of Meteorology
A. Gordon	The Flinders University of South Australia
A. Henderson-Sellers	MacQuarie University
A.C. Hirst	CSIRO (Division of Atmospheric Research)
B.G. Hunt	CSIRO (Division of Atmospheric Research)
M. Jones	MacQuarie University (currently at the Building Research Establishment, UK)
D. Karoly	Monash University
J. Katzfey	CSIRO (Division of Atmospheric Research)
D. Lee	Bureau of Meteorology
J. Leech	Royal Australian Navy Hydrographic Office
G. Lennon	National Tidal Facility
M.J. Manton	Bureau of Meteorology Research Centre

B. McAvaney	Bureau of Meteorology Research Centre
A.D. McEwan	CSIRO (Division of Oceanography)
J. McGregor	CSIRO (Division of Atmospheric Research)
C. Mitchell	CSIRO (Division of Atmospheric Research)
G. Morvell	Department of the Arts, Sport, the Environment & Territories
A. Moore	CSIRO (Division of Atmospheric Research)
N. Nicholls	Bureau of Meteorology Research Centre
G.I. Pearman	CSIRO (Division of Atmospheric Research)
A.B. Pittock	CSIRO (Division of Atmospheric Research)
C.M.R. Platt	CSIRO (Division of Atmospheric Research)
S.B. Power	Bureau of Meteorology Research Centre
P.G. Price	Bureau of Meteorology
L. Rikus	Bureau of Meteorology Research Centre
B. Ryan	CSIRO (Division of Atmospheric Research)
N. Smith	Bureau of Meteorology Research Centre
B. Stewart	Bureau of Meteorology
N. Streten	Bureau of Meteorology
J. Taylor	Australian National University
W.J.McG. Tegart	Australian Science and Technology Council
G.B. Tucker	CSIRO (Division of Atmospheric Research)
M. Voice	Bureau of Meteorology
J. Warne	Bureau of Meteorology
I. Watterson	CSIRO (Division of Atmospheric Research)
P.H. Whetton	CSIRO (Division of Atmospheric Research)
M. Williams	Monash University
N. Young	Australian Antarctic Division
J.W. Zillman	Bureau of Meteorology

BELGIUM

A. Berger	Institut d'Astronomie et de Géophysique
G.R. Demarée	Royal Meteorological Institute

BRAZIL

L.G. Meira Filho	Instituto Nacional de Pesquisas Espaciais

BURKINA FASO

G. Thiombiano	Institute Burkinabé de l'Energie

CAMEROON

T.O. Fouda	Directorate of Meteorology

CANADA

G.J. Boer	Canadian Climate Centre (currently at ECMWF, UK)
R.A. Clarke	Bedford Institute of Oceanography
K.L. Denman	Institute of Ocean Sciences
J.P. Hall	Forestry Canada
L.D. Harvey	University of Toronto
H. Hengeveld	Canadian Climate Centre
A.P. Jaques	Environment Canada
P. LeBlond	University of British Columbia
N. McIlveen	Energy, Mines and Resources Canada
G.T. Needler	Bedford Institute of Oceanography
R. Stoddart	Fisheries and Oceans Canada
J.M.R. Stone	Canadian Climate Centre
F. Zwiers	Canadian Climate Centre

CHAD
M. Guetti   Directorate of Meteorology

CHILE
J.M. Ovalle   Ministerio de Relaciones Exteriores

CHINA
L. Chen   State Meteorological Administration, China
Y. Ding   State Meteorological Administration, China
C. Fu   Institute of Atmospheric Physics
J. Luo   State Meteorological Administration, China
M.-X. Wang   Institute of Atmospheric Physics
S.-W. Wang   Beijing University
G. Wu   Institute of Atmospheric Physics
Y. Xu   Institute of Atmospheric Physics
D. Ye   Institute of Atmospheric Physics
Z. Yuan   Institute of Atmospheric Physics
Q.C. Zeng   Institute of Atmospheric Physics
X. Zhang   Institute of Atmospheric Physics

CONGO
E.A. Nyanga   Direction Générale de l'Environnement

CUBA
T. Gutiérrez Pérez   Institute of Meteorology

DENMARK
P. Frich   Danish Meteorological Institute
E. Friis-Christensen   Danish Meteorological Institute
K. Lassen   Danish Meteorological Institute
L. Laursen   Danish Meteorological Institute
L.P. Prahm   Danish Meteorological Institute

ETHIOPIA
L. Gonfa   National Meteorological Services Agency
T. Haile   National Meteorological Services Agency

FINLAND
E. Holopainen   University of Helsinki

FRANCE
G. Lambert   Centre des Faibles Radioactivités, CNRS

GERMANY
M.O. Andreae   Max-Planck Institute for Meteorology
L. Bengtsson   Max-Planck Institute for Meteorology
P. Crutzen   Max-Planck Institute for Chemistry
U. Cubasch   Max-Planck Institute for Meteorology
H. Flohn   Meteorological Institute
M. Heimann   Max-Planck Institute for Meteorology
K. Labitzke   Freie Universität Berlin
E. Maier-Reimer   Max-Planck Institute for Meteorology
B. Santer   Max-Planck Institute for Meteorology
C.-D. Schönwiese   Johann Wolfgang Goethe-Universität

GHANA
J.K. Danso   Environmental Protection Council

GUINEA
A.K. Bangoura   Ministry of Natural Resources and the Environment

GUYANA
W. Chin   Agency for Health Sciences, Education, Environment and Food Policy
S. Khan   Meteorological Service

INDIA
A.P. Mitra   National Physical Laboratory
G.B. Pant   Indian Institute of Tropical Meteorology
M. Sanwal   Ministry of Environment and Forests
S.K. Sinha   Indian Agricultural Research Centre

INDONESIA
A. Sugandhy   Ministry of Population and Environment

IRAN
M. Tohoder   Department of the Environment

ITALY
M. Conte   Italian Meteorological Service
A. Sutera   University of Camerino
S. Tibaldi   University of Bologna

JAPAN
A. Kitoh   Meteorological Research Institute
T. Nitta   Japan Meteorological Agency
T. Tokioka   Japan Meteorological Agency
T. Tsuyuki   Japan Meteorological Agency
T. Yasunari   University of Tsukuba

KENYA
J.K. Njihia   Kenya Meteorological Department
L.J. Ogallo   University of Nairobi

LIBYA
M.E. Alem   Meteorological Department
S.A. Younis   Meteorological Department

MALAYSIA
A.L. Chong   Meteorological Service

MALDIVES
H. Riza   Department of Meteorology

MALTA
J.M. Mifsud   Meteorological Office

MEXICO
A.C. Conde   Universidad Nacional Autonoma de Mexico

MONGOLIA
Mijiddorj   Hydrometeorological Research Institute

NETHERLANDS
A. Baede   Royal Netherlands Meteorological Institute (KNMI)
C. Schuurman   Royal Netherlands Meteorological Institute (KNMI)

NEW ZEALAND
B.J. Forde   Department of Scientific & Industrial Research
J.E. Hay   University of Auckland
M.R. Manning   Currently at NOAA/Department of Commerce in the USA
J. Salinger   New Zealand Meteorological Service
R.S. Whitney   Coal Research Association
D. Wratt   New Zealand Meteorological Service

NIGERIA
E.O. Oladipo   Ahmadu Bello University

NORWAY
B. Aune   Meteorological Office
K. Pedersen   University of Oslo
A. Rosland   State Pollution Control Authority

PAKISTAN
M. Hanif   Council for Scientific & Industrial Research

PAPUA NEW GUINEA
C. Kalwin   Currently at the South Pacific Regional Environment Programme, Western Samoa

PERU
E. Culqui Diaz   National Meteorological Service

PHILIPPINES
E.A. Gonzales   Department of Environment and Natural Resources

POLAND
B. Jakubiak   Institute of Meteorology and Water Management
E. Radwanski   Technical University Warsaw
M. Sadowski   Institute of Meteorology and Water Management

RUSSIA
G.S. Golitsyn   Academy of Sciences, Institute of Atmospheric Physics
I.L. Karol   Main Geophysical Observatory
K. Kondratyev   Academy of Sciences, Institute for Lake Research
V.P. Meleshko   Main Geophysical Observatory
K.Ya. Vinnikov   Main Geophysical Observatory (currently at GFDL, USA)

SAUDI ARABIA
A. Al-Dabbagh   King Fahd University of Petroleum and Minerals
M.A.A. Bin Afeef   King Abdul-Aziz University
A.M. Henaidi   Meteorology & Environmental Protection Administration
D.A. Olsen   Meteorology & Environmental Protection Administration
W.W. Sanford   Meteorology & Environmental Protection Administration
A. Wood   Meteorology & Environmental Protection Administration

SENEGAL
M. Seck   Ministry of Equipment, Transport and Sea

SIERRA LEONE
D.S. Lansama   Meteorological Office

SOLOMON ISLANDS
M. Ariki   Solomon Islands Meteorological Service

ST. LUCIA
N.C. Singh   Caribbean Environmental Health Institute

SUDAN
F. Elkhidir   Meteorological Department

SWEDEN
J. Heintzenberg   Stockholm University
H. Rodhe   Stockholm University

SWITZERLAND
W. Haeberli   Swiss Federal Institute of Technology
U. Siegenthaler   University of Bern

TANZANIA
B.S. Nyenzi   Directorate of Meteorology

THAILAND
S. Wangwongwatana   Office of National Environmental Board

TONGA
V. Fuavao   Currently at the South Pacific Regional Environment Programme, Western Samoa

TRINIDAD AND TOBAGO
L. Heileman   Carribean Meteorological Service

TUNISIA
C. Ben M'hamed   National Institute of Meteorology
M. Ketata   National Institute of Meteorology
Y. Labane   National Institute of Meteorology
K. Tounsi   National Institute of Meteorology

UGANDA
A.W. Majugu   Ministry of Water, Minerals & Environmental Protection

UNITED KINGDOM
A.J. Apling   Department of the Environment
D. Burdekin   Forestry Commission
D.J. Carson   Hadley Centre for Climate Research and Prediction, Meteorological Office
R.G. Derwent   Department of the Environment
D.J. Fisk   Department of the Environment
P.H. Freer-Smith   Forestry Commission Research Division
J. Grove   University of Cambridge
B.J. Hoskins   University of Reading
J.T. Houghton   Hadley Centre for Climate Research and Prediction, Meteorological Office
M. Hulme   University of East Anglia, Climatic Research Unit
P.G. Jarvis   University of Edinburgh
G.J. Jenkins   Met. Research Flight, Meteorological Office
P.D. Jones   University of East Anglia, Climatic Research Unit

R. Maryon	Meteorological Office
P.J. Mason	Meteorological Office
J.F.B. Mitchell	Hadley Centre for Climate Research and Prediction, Meteorological Office
D.E.Parker	Hadley Centre for Climate Research and Prediction, Meteorological Office
J.M. Penman	Department of the Environment
P.R. Rowntree	Hadley Centre for Climate Research and Prediction, Meteorological Office
K. Shine	University of Reading
A. Slingo	Hadley Centre for Climate Research and Prediction, Meteorological Office
F.B. Smith	Meteorological Office
F.A. Street-Perrott	University of Oxford
R.A. Warrick	Climatic Research Unit, University of East Anglia
D.A. Warrilow	Department of the Environment
D.J. Webb	Institute of Oceanographic Sciences
T.M.L. Wigley	Climatic Research Unit, University of East Anglia

UNITED STATES OF AMERICA

D.L. Albritton	NOAA Aeronomy Laboratory
J. Angell	NOAA Air Resources Laboratory
P. Arkin	NOAA National Meteorological Center
J. Ausubel	Rockefeller University
R. Balling	Arizona State University
T.P. Barnett	Scripps Institute of Oceanography
S. Barr	Los Alamos National Laboratory
E. Barron	Pennsylvania State University
R.G. Barry	University of Colorado
L. Beck	USEPA Air and Energy Research Laboratory
P. Beedlow	Environmental Research Laboratory at Corvallis
C. Bernabo	Science and Policy Association
D. Blake	University of California at Irvine
J.W. Bozzelli	New Jersey Institute of Technology
A.J. Broccoli	NOAA/US Department of Commerce
S. Brown	University of Illinois
O. Brown	University of Miami
W.S. Broecker	Lamont-Doherty Geological Observatory
K. Bryan	NOAA Geophysical Fluid Dynamics Laboratory
M. Chahine	NASA Jet Propulsion Laboratory
E. Christian	US Geological Survey
J. Christy	University of Alabama
R. Cicerone	University of California at Irvine
P. Connell	Lawrence Livermore National Laboratory
R.C. Dahlman	US Department of Energy
R.E. Dickinson	University of Arizona
J. Dignon	Lawrence Livermore National Laboratory
B.G. Drake	Smithsonian Environmental Research Center
J. Edmonds	Battelle Pacific N.W. Laboratory
W.P. Elliott	NOAA Air Resources Laboratory
H.W. Ellsaesser	Lawrence Livermore National Laboratory
C. Elvidge	US Environmental Protection Agency
W.R. Emanuel	Oak Ridge National Laboratory
J. Fein	National Science Foundation
M. Fernau	Argonne National Laboratory
J.W. Firor	National Center for Atmospheric Research
M.M. Fogel	University of Arizona
I. Fung	NASA Goddard Space Flight Center

H.L. Gholz	University of Florida
I. Goklany	US Department of the Interior
A. Gruber	NOAA/NESDIS
H. Gruenspecht	US Department of Energy
R.P. Hangebrauck	USEPA Air and Energy Research Laboratory
K. Hogan	US Environmental Protection Agency
R.A. Houghton	Woods Hole Research Center
A. Janetos	NASA Headquarters
S. Kane	NOAA Economics Group
T. Karl	NOAA National Climate Data Center
J. Kelmelis	Department of the Interior/USGS
M.A.K. Khalil	Oregon Graduate Institute of Science & Technology
S.J.S. Khalsa	University of Colorado
D.A. Kirchgessner	USEPA Air and Energy Research Laboratory
M. Ko	Atmospheric and Environmental Research, Inc.
G. Kukla	Columbia University
C.B. Leovy	University of Washington
J.S. Levine	NASA Langley Research Center
R.S. Lindzen	Massachusetts Institute of Technology
J. Logan	Harvard University
A. Lugo	Institute of Tropical Forestry
M.C. MacCracken	Lawrence Livermore National Laboratory
W. MacPost	Oak Ridge National Laboratory
J. Mahlman	NOAA Geophysical Fluid Dynamics Laboratory
S. Manabe	NOAA Geophysical Fluid Dynamics Laboratory
G. Marland	Oak Ridge National Laboratory
G.A. Meehl	National Center for Atmospheric Research
P.J. Michaels	University of Virginia
P. Milke	Colorado State University
C. Milly	US Geological Survey
S.C. Morris	Brookhaven National Laboratory
L. Mulkey	Environmental Research Laboratory at Athens
P.J. Mutschlecner	Los Alamos National Laboratory
R. Newell	Massachusetts Institute of Technology
L. Newman	Brookhaven National Laboratory
G. Ohring	NOAA/NESDIS
R. Oremland	US Geological Survey
C.L. Parkinson	NASA Goddard Space Flight Center
T.H. Peng	Oak Ridge National Laboratory
J. Pershing	US Department of State
L. Pettinger	US Geological Survey
R. Pielke	Colorado State University
J.S. Pinto	US Environmental Protection Agency
R. Poore	US Geological Survey
W.M. Post	Oak Ridge National Laboratory
M.J. Prather	NASA Goddard Institute of Space Studies (now at University of California, Irvine)
R.G. Prinn	Massachusetts Institute of Technology
C. Purvis	USEPA Air and Energy Research Laboratory
P. Quay	University of Washington
R. Quayle	National Climatic Data Center
S. Ragone	US Geological Survey
V. Ramanathan	University of California at San Diego
D.G. Randall	Colorado State University
A.R. Ravishankara	NOAA Aeronomy Laboratory

G. Reid	NOAA Environmental Research Laboratory
W. Rhodes	USEPA Air and Energy Research Laboratory
M.R. Riches	US Department of Energy
P. Ringold	US Environmental Protection Agency
C. Riordan	US Environmental Protection Agency
R.L. Ritschard	Lawrence Berkeley Laboratory
E.S. Sarachik	University of Washington
J.L. Sarmiento	Princeton University
R.A. Schiffer	NASA Headquarters
D. Schimel	Colorado State University
M.E. Schlesinger	University of Illinois
S.E. Schwartz	Brookhaven National Laboratory
P. Sellars	NASA Goddard Space Flight Center
J. Shannon	Argonne National Laboratory
J. Shukla	University of Maryland
L. Smith	US Environmental Protection Agency
S. Solomon	NOAA Aeronomy Laboratory
J.R. Spradley	NOAA Headquarters
D.W. Stahle	University of Arkansas
G.L. Stephens	Colorado State University
R. Stolarski	NASA Goddard Space Flight Center
R. Stouffer	NOAA Geophysical Fluid Dynamics Laboratory
L.L. Stowe	NOAA/NESDIS
D.G. Streets	Argonne National Laboratory
A.E. Strong	NOAA/NESDIS
E. Sundquist	US Geological Survey
N.-D. Sze	Atmospheric and Environmental Research, Inc.
L. Talley	Scripps Institute of Oceanography
P. Tans	NOAA Climate Monitoring and Diagnostics Laboratory
J. Theon	NASA Headquarters
M.H. Thiemens	University of California at San Diego
S. Thorneloe	USEPA Air and Energy Research Laboratory
D. Tingey	Environmental Research Laboratory at Corvallis
D. Tirpak	US Environmental Protection Agency
K.E. Trenberth	National Center for Atmospheric Research
H. Virgi	National Science Foundation
J. Vitko, Jr.	Sandia National Laboratory
D.W.R. Wallace	Brookhaven National Laboratory
W.-C. Wang	State University of New York at Albany
W.L. Warnick	US Department of Energy
W. Washington	National Center for Atmospheric Research
H. Watson	US Department of the Interior
R.T. Watson	NASA Headquarters
J. Wiener	US Department of Justice
E.R. Williams	US Department of Energy
D. Winstanley	US Department of Energy
S.C. Wofsy	Harvard University
F.B. Wood	US Congress Office of Technology Assessment
S. Woodruff	NOAA Environmental Research Laboratories
R.C. Worrest	US Environmental Protection Agency
D.J. Wuebbles	Lawrence Livermore National Laboratory

VANUATU

H. Tazsilas	Currently at the South Pacific Regional Environment Programme, Western Samoa

VENEZUELA
N. Pereira   Ministry of Energy and Mines
E. Sanhueza   Instituto Venezolano de Investigaciones Cientificas

VIETNAM
N. Trong Hieu   Hydrometeorological Service

WESTERN SAMOA
P. Holthus   South Pacific Regional Environment Programme

ZIMBABWE
A. Makarau   Department of Meteorological Services

UNITED NATIONS SPECIALIZED AGENCIES
W.G. Sombroek   Food and Agriculture Organization
J. Van de Vaate   International Atomic Energy Agency
P.F. Schwengels   OECD (Environment Directorate)
P. Usher   United Nations Environment Programme (GEMS)
R. Newson   World Meteorological Organization (WCRP)

NON-GOVERNMENTAL ORGANIZATIONS
A. McCulloch   Alternative Fluorocarbon Environmental Acceptability Study/ICI C&P Ltd, UK
D.M. Hughes   Australian Coal Association
I. Hughes   British Coal Corporation
P.W. Sage   British Coal Corporation
D.H. Pearlman   The Climate Council, USA
M. McFarland   Du Pont Chemicals, USA
R.A. Beck   Edison Electric Institute, USA
M. Oppenheimer   Environmental Defense Fund, USA
J. Leggett   Greenpeace International
C.D. Holmes   National Coal Association, USA
D. Lashof   Natural Resources Defence Council, USA
A. Markham   World Wide Fund for Nature International
K. Gregory   World Coal Institute/British Coal Corporation
M. Jefferson   World Energy Council

Appendix 4

ACRONYMS

AFEAS	Alternative Fluorocarbon Environmental Acceptability Study
AGCM	Atmosphere General Circulation Model
ALE/GAGE	Atmospheric Lifetime Experiment/Global Atmospheric Gases Experiment
AMIP	Atmospheric Model Intercomparison Project
ASF	Atmospheric Stabilization Framework
AVHRR	Advanced Very High Resolution Radiometer
BMRC	Bureau of Meteorology Research Centre, Australia
CAS	Commission for Atmospheric Sciences
CCC	Canadian Climate Centre
CCN	Cloud Condensation Nuclei
CEC	Commission of the European Communities
CHEMRAWN	Chemical Research Applied to World Needs
CIES	Climatic Impacts on the Environment and Society (Conference in Tsukuba, Japan, 1991)
CLW	Cloud Liquid Water
CNRM	Centre National de Recherches Météorologiques, France
CNRS	Centre National de la Recherche Scientifique, France
COADS	Comprehensive Ocean Air Data Set (USA)
CONCAWE	Conservation of Clean Air and Water, Europe (an oil companies organization)
CRU	Climatic Research Unit, University of East Anglia, UK
CSIRO	Commonwealth Scientific & Industrial Research Organisation, Australia
CW	Cloud Water
DJF	December-January-February
DSIR	Department of Scientific and Industrial Research, New Zealand
ECHAM	European Centre/Hamburg Model (ECMWF/MPI)
ECMWF	European Centre for Medium-Range Weather Forecasts, UK
EDF	Environmental Defense Fund, USA
EIS	Energy and Industry Sub-Group (of IPCC WG3)
ENSO	El Niño-Southern Oscillation
EOF	Empirical Orthogonal Function
EOS	Earth Observing System
EPA	Environmental Protection Agency, USA

ERBE	Earth Radiation Budget Experiment
ERL	Environmental Research Laboratory (NOAA), USA
FAO	Food and Agriculture Organization (of the UN)
FC	Fixed Cloud
FRAM	Fine Resolution Antarctic Model
GARP	Global Atmospheric Research Programme
GCM	General Circulation Model
GCOS	Global Climate Observing System
GDP	Gross Domestic Product
GEF	Global Environment Facility
GEMS	Global Environmental Monitoring System
GEWEX	Global Energy and Water Cycle Experiment
GFDL	Geophysical Fluid Dynamics Laboratory, USA
GISS	Goddard Institute of Space Sciences (NASA), USA
GNP	Gross National Product
GOSTA	Global Ocean Surface Temperature Atlas
GPP	Gross Primary Production
GSFC	Goddard Space Flight Center (NASA), USA
GWP	Global Warming Potential
IAEA	International Atomic Energy Agency
IAHS	International Association of Hydrological Sciences
ICRCCM	Intercomparison of Radiation Codes in Climate Models
ICSU	International Council of Scientific Unions
IEA	International Energy Agency
IGBP	International Geosphere-Biosphere Programme
INPE	Instituto Nacional de Pesquisas Espaciais, Brazil
IOC	International Ozone Commission
IOS	Institute of Oceanographic Sciences, UK
IPCC	Intergovernmental Panel on Climate Change
ISCCP	International Satellite Cloud Climatology Project
IS92	IPCC Scenarios 1992
IVIC	Instituto Venezolano de Investigaciones Cientificas, Venezuela
JJA	June-July-August
JMA	Japan Meteorological Agency
JSC	Joint Scientific Committee (of WMO and ICSU)
KNMI	Royal Netherlands Meteorological Institute
LLNL	Lawrence Livermore National Laboratory, USA
LMD	Laboratoire de Météorologie Dynamique du CNRS
LRTAP	Long-Range Transboundary Air Pollution
LWC	Liquid Water Content
MCA	Moist Convective Adjustment
MDB	Marine Data Bank (UKMO)
MEPA	Meteorological & Environmental Protection Administration, Saudi Arabia
NESDIS	National Environmental Satellite Data & Information Service (NOAA), USA
MGO	Main Geophysical Observatory, Russia
MIT	Massachusetts Institute of Technology, USA
MONEG	Monsoon Numerical Experimentation Group

MPI	Max-Planck Institute, Germany
MRI	Meteorological Research Institute, Japan
MSU	Microwave Sounding Unit

NACD	North Atlantic Climatological Dataset
NASA	National Aeronautics and Space Administration, USA
NATO	North Atlantic Treaty Organisation
NCAR	National Center for Atmospheric Research, USA
NERC	Natural Environment Research Council, UK
NGL	Natural Gas Liquids
NH	Northern Hemisphere
NMC	National Meteorological Center, USA
NOAA	National Oceanic and Atmospheric Administration, USA
NRDC	Natural Resources Defense Council, USA
NSF	National Science Foundation, USA
NWP	Numerical Weather Prediction

1-D	One Dimensional (also 2-D and 3-D)
ODP	Ozone Depletion Potential
OECD	Organization for Economic Cooperation and Development
OGCM	Ocean General Circulation Model
OLR	Outgoing Longwave Radiation

PC	Penetrative Convection
PCMDI	Program for Climate Model Diagnosis and Intercomparison (at LLNL, USA)
PMIP	Palaeoclimatic Model Intercomparison Project
PW	Precipitable Water

RCM	Radiative Convective Model
RCW	Rapidly Changing World
RH	Relative Humidity
RSWG	Response Strategies Working Group (IPCC WG3)

SA90	IPCC Scenario A 1990
SAGE	Stratospheric Aerosol and Gas Experiment
SCW	Slowly Changing World
SCOPE	Scientific Committee On Problems of the Environment
SH	Southern Hemisphere
SMA	State Meteorological Administration, China
SORG	Stratospheric Ozone Review Group, UK
SPCZ	South Pacific Convergence Zone
SST	Sea Surface Temperature
SUNY	State University of New York, USA

TIROS	Television and Infrared Observation Satellite
TOGA	Tropical Ocean and Global Atmosphere
TOMS	Total Ozone Mapping Spectrometer
TOA	Top of the Atmosphere
TPER	Total Primary Energy Requirement

UARS	Upper Atmosphere Research Satellite
UCAR	University Corporation for Atmospheric Research
UKMO	Meteorological Office, UK
UN	United Nations

UNEP	United Nations Environment Programme
UNESCO	United Nations Educational, Scientific and Cultural Organization
UV	Ultraviolet
VOC	Volatile Organic Compounds
WCI	World Coal Institute
WCRP	World Climate Research Programme
WEC	World Energy Council
WGI	Working Group 1 (of IPCC)
WGNE	Working Group on Numerical Experimentation
WMO	World Meteorological Organization
WOCE	World Ocean Circulation Experiment
WRI	World Resources Institute
WWF	World Wide Fund for Nature

Appendix 5

UNITS

SI (Systeme Internationale) Units:

Physical Quantity	Name of Unit	Symbol
length	metre	m
mass	kilogram	kg
time	second	s
thermodynamic temperature	kelvin	K
amount of substance	mole	mol

Fraction	Prefix	Symbol	Multiple	Prefix	Symbol
10^{-1}	deci	d	10	deca	da
10^{-2}	centi	c	10^2	hecto	h
10^{-3}	milli	m	10^3	kilo	k
10^{-6}	micro	μ	10^6	mega	M
10^{-9}	nano	n	10^9	giga	G
10^{-12}	pico	p	10^{12}	tera	T
10^{-15}	femto	f	10^{15}	peta	P
10^{-18}	atto	a			

Special Names and Symbols for Certain SI-Derived Units:

Physical Quantity	Name of SI Unit	Symbol for SI Unit	Definition of Unit
force	newton	N	$kg\ m\ s^{-2}$
pressure	pascal	Pa	$kg\ m^{-1}s^{-2}(=Nm^{-2})$
energy	joule	J	$kg\ m^2\ s^{-2}$
power	watt	W	$kg\ m^2s^{-3}(=Js^{-1})$
frequency	hertz	Hz	s^{-1}(cycle per second)

Decimal Fractions and Multiples of SI Units Having Special Names:

Physical Quantity	Name of Unit	Symbol for Unit	Definition of Unit
length	ångstrom	Å	$10^{-10}\ m = 10^{-8}cm$
length	micrometre	μm	$10^{-6}m = \mu m$
area	hectare	ha	$10^4\ m^2$
force	dyne	dyn	$10^{-5}\ N$
pressure	bar	bar	$10^5\ N\ m^{-2}$
pressure	millibar	mb	$1hPa$
weight	tonne	t	$10^3\ kg$

Non- SI Units:

°C	degrees Celsius (0°C = 273K approximately)
	Temperature differences are also given in °C (=K) rather than the more correct form of "Celsius degrees".
ppmv	parts per million (10^6)by volume
ppbv	parts per billion (10^9) by volume
pptv	parts per trillion (10^{12}) by volume
bp	(years) before present
kpb	thousands of years before present
mbp	millions of years before present

The units of mass adopted in this report are generally those which have come into common usage, and have deliberately not been harmonized, e.g.,

kt	kilotonnes
GtC	gigatonnes of carbon (1 GtC = 3.7 Gt carbon dioxide)
MtN	megatonnes of nitrogen
TgC	teragrams of carbon
TgN	teragrams of nitrogen
TgS	teragrams of sulphur

Appendix 6

CHEMICAL SYMBOLS

O	atomic oxygen		C_2H_6	ethane
O_2	molecular oxygen		C_3H_8	propane
O_3	ozone		C_2H_4	ethylene
N	atomic nitrogen		C_2H_2	acetylene
N_2	molecular nitrogen		CH_3Br	methyl bromide
N_2O	nitrous oxide		CH_3Cl	methyl chloride
NO	nitric oxide		CH_3CCl_3	methyl chloroform
NO_2	nitrogen dioxide		CH_2Cl_2	dichloromethane
NO_3	nitrate radical		$CHCl_3$	chloroform, trichloromethane
NO_y	total active nitrogen		Br	bromine
NO_x	nitrogen oxide		BrO	bromine monoxide
HNO_3	nitric acid		F	fluorine
NH_3	ammonia		PAN: $CH_3CO_3NO_2$	peroxyacetylnitrate
H	hydrogen		pCO_2	partial pressure CO_2
H_2O	water		CFC	chlorofluorocarbon
H_2O_2	hydrogen peroxide		HC	hydrocarbon
OH	hydroxyl		HCFC	hydrochlorofluorocarbon
HO_2	hydroperoxyl		HFC	hydrofluorocarbon
C	carbon		NMHC	non-methane hydrocarbons
CO	carbon monoxide		VOC	volatile organic compound
CO_2	carbon dioxide		DMS	dimethylsulphide
CS_2	carbon disulphide			
COS	carbonyl sulphide		CFC-11	$CFCl_3$, CCl_3F
CCl_4	carbon tetrachloride			(trichlorofluoromethane)
S	sulphur		CFC-12	CF_2Cl_2, CCl_2F_2
SO_2	sulphur dioxide			(dichlorodifluoromethane)
SO_x	sulphur oxide		CFC-13	CF_3Cl, $CClF_3$
SF_6	sulphur hexafluoride			(chlorotrifluoromethane)
H_2SO_4	sulphuric acid		CFC-14	CF_4
H_2S	hydrogen sulphide			(tetrafluoromethane)
HCl	hydrochloric acid		CFC-113	$C_2F_3Cl_3$, CCl_2FCClF_2
HOCl	hypochlorous acid			(trichlorotrifluoroethane)
Cl	chlorine		CFC-114	$C_2F_4Cl_2$, $CClF_2CClF_2$
Cl_2	molecular chlorine			(dichlorotetrafluoroethane)
ClO	chlorine monoxide		CFC-115	C_2F_5Cl, $CClF_2CF_3$
ClO_2	chlorine dioxide			(chloropentafluoroethane)
$ClONO_2$	chlorine nitrate		CFC-116	CF_3CF_3, C_2F_6
CH_4	methane			(hexafluoroethane)

HCFC-22	CHF_2Cl
	(chlorodifluoromethane)
HCFC-123	$C_2HF_3Cl_2$ ($CHCl_2CF_3$)
HCFC-124	$CHFClCF_3$
HFC-125	CHF_2CF_3
HCFC-132b	$C_2H_2F_2Cl_2$
HFC-134a	$C_2H_2F_4$ (CH_2FCF_3)
HCFC-141b	CH_3CFCl_2
HCFC-142b	CH_3CF_2Cl
HFC-143a	CH_3CF_3
HFC-152a	CH_3CHF_2
HALON 1211	CF_2BrCl ($CBrClF_2$)
	(bromodichloromethane)
HALON 1301	CF_3Br ($CBrF_3$)
	(bromotrifluoromethane)
HALON 2402	$C_2F_4Br_2$
	(dibromotetrafluoroethane)